Lecture Notes in Mathematics   1535

Editors:
A. Dold, Heidelberg
B. Eckmann, Zürich
F. Takens, Groningen

Anthony B. Evans

Orthomorphism Graphs of Groups

Springer-Verlag

Berlin Heidelberg New York
London Paris Tokyo
Hong Kong Barcelona
Budapest

Author

Anthony B. Evans
Department of Mathematics and Statistics
Wright State University
Dayton, OH 45435, USA

Mathematics Subject Classification (1991): Primary: 05-02, 05B15
                                             Secondary: 51E14, 51E15

ISBN 3-540-56351-2 Springer-Verlag Berlin Heidelberg New York
ISBN 0-387-56351-2 Springer-Verlag New York Berlin Heidelberg

© Springer-Verlag Berlin Heidelberg 1992
Printed in Germany

Typesetting: Camera ready by author
46/3140-543210 - Printed on acid-free paper

Preface

The study of orthomorphism graphs of groups has its origin in the use of orthomorphisms and complete mappings to construct mutually orthogonal sets of Latin squares. To help other mathematicians who wish to work in this area, a reference work is needed. None exists at present and work on the subject is scattered throughout the literature, often in a form that does not suggest any connection to orthomorphisms.

In writing this monograph I have tried to do more than survey work done so far. In this monograph I have attempted to consolidate known results and applications, to create a unified body of knowledge and to provide other mathematicians with the tools needed to work in this area. I have tried to lay down the beginnings of a framework for the theory of orthomorphism graphs of groups and their applications, incorporating topics from algebra and geometry into this theory. As one of the hopes for this project was that it would stimulate research in this area, I have suggested many problems and directions for future research in this field, which should provide algebraists and geometers as well as other researchers in combinatorics with questions to work on.

The material in this book should be accessible to any graduate student who has taken courses in group theory and field theory.

I would like to thank Wright State University for its support during the writing of this manuscript, part of which was written while on sabbatical, and my colleagues Manley Perkel, who worked with me in generating orthomorphisms using Cayley, and Terry McKee, who read part of the manuscript, and the referees for their helpful suggestions.

Contents

Chapter 6: Data for small groups

Chapter 7: Research directions

Chapter 1: Introduction.

Section 1.   Definitions and elementary results.

If G is a finite group we shall call any mapping $G \to G$ that permutes the elements of G a *permutation* of G. For a finite group G, written multiplicatively with identity e, let θ be a permutation of G fixing e. Then θ is an *orthomorphism* of G if the mapping $x \to x^{-1}\theta(x)$ is a permutation of G, and a *complete mapping* of G if the mapping $x \to x\theta(x)$ is a permutation of G. Orthomorphisms and complete mappings are closely related as θ is an orthomorphism of G if and only if the mapping $x \to x^{-1}\theta(x)$ is a complete mapping of G and a complete mapping of G if and only if the mapping $x \to x\theta(x)$ is an orthomorphism of G.

Complete mappings were first defined and studied by Mann (1942) and the term orthomorphism was first used by Johnson, Dulmage and Mendelsohn (1961). While other terminologies have been used the most common alternative for orthomorphism has been orthogonal mapping, a term first used by Bose, Chakravarti, and Knuth (1960).

In this book we are concerned with mutually adjacent sets of orthomorphisms. These are used principally in the construction of mutually orthogonal Latin squares, nets, transversal designs, affine and projective planes, difference matrices, and generalized Hadamard matrices. Historically these are among the first applications found for orthomorphisms. Many applications of individual orthomorphisms have since been discovered. The intimate connection between neofields and near orthomorphisms (a generalization of orthomorphisms) will be described in section 4 of this chapter. For applications to the construction of triple systems, Mendelsohn designs, Room squares, and group sequencings, the interested reader should consult the papers in the reprint volumes edited by Hsu (1985 & 1987). These volumes also contain several papers dealing with neofields. Dénes and Keedwell's (1991) book on Latin squares also contains material on complete mappings, neofields, and group sequencings. One special type of orthomorphism deserves mention as it is the probably the most used in these constructions: A *starter* is an orthomorphism θ for which $\theta^{-1} = \theta$. A recent survey article by Dinitz and Stinson (to appear) contains much material on the use of starters in the construction of Room squares.

Let $\theta, \phi: G \to G$. Then θ is *adjacent* to ϕ, written $\theta \sim \phi$, if the mapping $x \to \theta(x)^{-1}\phi(x)$ is a permutation of G. Clearly $\theta \sim \phi$ if and only if $\phi \sim \theta$. Now $\theta: G \to G$ is a permutation if adjacent to the mapping $x \to e$, and if $\theta(e) = e$ then θ is an orthomorphism if adjacent to the mappings $x \to e$ and $x \to x$, and a complete mapping if adjacent to the mappings $x \to e$ and $x \to x^{-1}$.

While we could have left out the condition that $\theta(e) = e$ in defining orthomorphisms and complete mappings, there is little to be gained by doing so. To see this, suppose that we do not require orthomorphisms to fix the identity and for each mapping $\theta: G \to G$ and each

$a \in G$ we define the mapping $\theta_a: G \to G$ by $\theta_a(x) = \theta(x)a$. Then it is easy to check that θ is an orthomorphism if and only θ_a is an orthomorphism, θ is a complete mapping if and only if θ_a is a complete mapping, θ is adjacent to ϕ if and only if θ_a is adjacent to ϕ_b, and θ_a cannot be adjacent to θ_b. A simple characterization of adjacency of permutations follows.

Theorem 1.1. Let θ and ϕ be permutations of G. Then θ is adjacent to ϕ if and only if the mapping $x \to x^{-1}\theta\phi^{-1}(x)$ is a permutation of G.

Proof. The mapping $x \to x^{-1}\theta\phi^{-1}(x)$ is a permutation of G if and only if the mapping $y = \phi^{-1}(x) \to x = \phi(y) \to \phi(y)^{-1}\theta(y)$ is a permutation of G if and only if θ is adjacent to ϕ. ∎

As an easy corollary we obtain a characterization of adjacency of orthomorphisms.

Corollary 1.1. Let θ and ϕ be orthomorphisms of G. Then θ is adjacent to ϕ if and only if $\theta\phi^{-1}$ is an orthomorphism of G.

The *orthomorphism graph* of G has as its vertex set the orthomorphisms of G, adjacency being defined as above. An *orthomorphism graph* of G is any induced subgraph of the orthomorphism graph of G. We shall use the symbol Orth(G) to denote both the set of orthomorphisms of G and the orthomorphism graph of G. Orthomorphism graphs were first studied by Johnson, Dulmage, and Mendelsohn (1961).

We shall borrow terminology from graph theory. In particular, if $\theta \in \mathcal{H}$, an orthomorphism graph of G, then the *degree* of θ in $\mathcal{H}$ is the number of orthomorphisms in $\mathcal{H}$ that are adjacent to θ. Any orthomorphism (in $\mathcal{H}$), adjacent to θ, will be called a *neighbor* of θ (in $\mathcal{H}$). The *degree* of an orthomorphism of G is its degree in Orth(G). An *r-clique* of an orthomorphism graph $\mathcal{H}$ of G is a set of r mutually adjacent orthomorphisms in $\mathcal{H}$. The *clique number* of $\mathcal{H}$, denoted $\omega(\mathcal{H})$, is the largest value of r for which $\mathcal{H}$ admits an r-clique.

Example 1.1. Let $\mathcal{P}(G) = \{\phi_r : \phi_r(x) = x^r, \phi_r \in \text{Orth}(G)\}$. Now ϕ_r is a permutation of G if and only if r is relatively prime to $|G|$. Hence $\phi_r \in \mathcal{P}(G)$ if and only if r and $r - 1$ are both relatively prime to $|G|$ and $\phi_r \sim \phi_s$ if and only if $r - s$ is relatively prime to $|G|$. Thus if p is the smallest prime divisor of $|G|$ then the set $\{\phi_i : i = 2, \ldots, p - 1\}$ is a $(p - 2)$-clique of $\mathcal{P}(G)$. Using the pigeonhole principle it is easy to prove that no larger clique of $\mathcal{P}(G)$ can be constructed and so $\omega(\mathcal{P}(G)) = p - 2$.

What is the value of $\omega(\text{Orth}(G))$? This question has been answered for very few classes of groups. One easily obtained upper bound is the following.

Theorem 1.2. If G is non-trivial then $\omega(\text{Orth}(G)) \leq |G| - 2$.

Proof. Let $\theta_1, \ldots, \theta_r$ be an r-clique of Orth(G). If $a \neq e$ then $\theta_1(a), \ldots, \theta_r(a)$ are all distinct and equal to neither e nor a. Hence $r \leq |G| - 2$. ∎

In view of this result we will define a *complete set of orthomorphisms of G* to be an $(|G| - 2)$ - clique of Orth(G). Does any group admit a complete set of orthomorphisms?

Example 1.2. Let G be an elementary abelian group. We may think of G as the additive group of a finite field. Then the mapping $x \to ax$ will be an orthomorphism of G if and only if $a \neq 0, 1$ and two such mappings $x \to ax$ and $x \to bx$ will be adjacent if and only if $a \neq b$. Thus the set $\{x \to ax : a \neq 0, 1\}$ is a complete set of orthomorphisms of G.

The elementary abelian groups are the only groups known to admit complete sets of orthomorphisms. Are there other groups that admit complete sets of orthomorphisms? In particular, are there groups of non-prime power order that admit complete sets of orthomorphisms? The latter question is important as it relates to the question of the existence of projective (equivalently affine) planes of non-prime power order. Such planes have long been conjectured not to exist.

We now give a direct product construction for orthomorphisms. This construction was given in difference matrix form by Jungnickel (1978). This yields lower bounds for the clique numbers of orthomorphism graphs. These bounds, though weak, are the best lower bounds presently available for many groups.

Theorem 1.3. Let $\theta \in$ Orth(G) and $\theta' \in$ Orth(G') and define $\theta \times \theta' : G \times G' \to G \times G'$ by $\theta \times \theta'(x, x') = (\theta(x), \theta'(x'))$. Then $\theta \times \theta' \in$ Orth($G \times G'$). Further, if $\theta, \phi \in$ Orth(G) and $\theta', \phi' \in$ Orth(G') then $\theta \times \theta'$ is adjacent to $\phi \times \phi'$ if and only if $\theta \sim \phi$ and $\theta' \sim \phi'$.
Proof. Routine calculation. ∎

The following are immediate corollaries.

Corollary 1.2. $\omega(\text{Orth}(G \times G')) \geq \text{Min}\{\omega(\text{Orth}(G)), \omega(\text{Orth}(G'))\}$.

Corollary 1.3. $\omega(\text{Orth}(G_1 \times \ldots \times G_n)) \geq \text{Min}\{\omega(\text{Orth}(G_1)), \ldots, \omega(\text{Orth}(G_n))\}$.

We shall consider two types of mappings Orth(G) → Orth(G) that are important in studying the structure of orthomorphism graphs. If θ is a permutation of G and T maps permutations of G into permutations of G then we shall use T[θ] to denote the image of θ under T. A bijection T: Orth(G) → Orth(G) is an *automorphism* of Orth(G) if T[θ] ~ T[ϕ] if and only if $\theta \sim \phi$ and a *congruence* of Orth(G) if the neighborhood of T[θ] is isomorphic to the neighborhood of θ. In general we cannot determine the full automorphism group or the full congruence group of Orth(G) without knowing completely the structure of Orth(G). We do however know some classes of automorphisms and congruences of orthomorphism graphs.

Let us define the following mappings from Orth(G) into Orth(G).

(i) H_α is defined by $H_\alpha[\theta] = \alpha\theta\alpha^{-1}$, $\alpha \in \text{Aut}(G)$.

(ii) T_g is defined by $T_g[\theta](x) = \theta(xg)\theta(g)^{-1}$.

(iii) R is defined by $R[\theta](x) = x\theta(x^{-1})$.

(iv) I is defined by $I[\theta](x) = \theta^{-1}(x)$.

We will call H_α an *homology*, T_g a *translation*, R a *reflection*, and I an *inversion*. These mappings and their relations were described implicitly in Singer (1960), for cyclic groups. The definitions of translations, homologies, and inversions used here are those of Johnson, Dulmage, and Mendelsohn (1961), though they used different notation and terminology. Chang, Hsiang, and Tai (1964) defined reflections for non-abelian groups.

Theorem 1.4. H_α, T_g and R are automorphisms of Orth(G) and I is a congruence of Orth(G).

Proof. If θ is an orthomorphism of G then $H_\alpha[\theta]$, $T_g[\theta]$, $R[\theta]$, and $I[\theta]$ are all permutations fixing e. To show that these are also orthomorphisms of G we use $\eta(x)$ to denote $x^{-1}\theta(x)$ and note that $x^{-1}H_\alpha[\theta](x) = x^{-1}\alpha\theta\alpha^{-1}(x) = \alpha(\alpha^{-1}(x^{-1})\theta(\alpha^{-1}(x))) = \alpha\eta\alpha^{-1}(x)$, $x^{-1}T_g[\theta](x) = x^{-1}\theta(xg)\theta(g)^{-1} = g\eta(xg)\theta(g)^{-1}$, $x^{-1}R[\theta](x) = x^{-1}x\theta(x^{-1}) = \theta(x^{-1})$, and $x^{-1}I[\theta](x) = x^{-1}\theta^{-1}(x) = (\eta\theta^{-1}(x))^{-1}$. These are clearly permutations.

To show that H_α is an automorphism of Orth(G) note that, by Corollary 1.1, $H_\alpha[\theta] \sim H_\alpha[\phi]$ if and only if $\alpha\theta\alpha^{-1}(\alpha\phi\alpha^{-1})^{-1} = \alpha\theta\phi^{-1}\alpha^{-1}$ is an orthomorphism if and only if $\theta\phi^{-1}$ is an orthomorphism if and only if $\theta \sim \phi$. We shall use $\delta(x)$ to denote $\theta(x)^{-1}\phi(x)$. To show that T_g is an automorphism of Orth(G) note that $T_g[\theta](x)^{-1}T_g[\phi](x) = \theta(g)\delta(xg)\phi(g)^{-1}$, which is a permutation if and only if $\theta \sim \phi$. To show that R is an automorphism of Orth(G) note that $R[\theta](x)^{-1}R[\phi](x) = \theta(x^{-1})^{-1}x^{-1}x\phi(x^{-1}) = \delta(x^{-1})$, which is a permutation if and only if $\theta \sim \phi$.

To show that I is a congruence note that if $\phi_1, \dots, \phi_r$ are the neighbors of θ then, by Corollary 1.1, $\phi_1\theta^{-1}, \dots, \phi_r\theta^{-1}$ are orthomorphisms and are neighbors of θ^{-1} as $(\phi_i\theta^{-1})(\theta^{-1})^{-1} = \phi_i$ is an orthomorphism for $i = 1, \dots, r$. Further, as $(\phi_i\theta^{-1})(\phi_j\theta^{-1})^{-1} = \phi_i\phi_j^{-1}$, it follows that $\phi_i\theta^{-1} \sim \phi_j\theta^{-1}$ if and only if $\phi_i \sim \phi_j$. Thus the neighborhood of θ is isomorphic to an induced subgraph of the neighborhood of $I[\theta]$. Similarly the neighborhood of $I[\theta]$ is isomorphic to an induced subgraph of the neighborhood of θ, and so the neighborhood of θ is isomorphic to the neighborhood of $I[\theta]$. ∎

Note that in general I is not an automorphism of Orth(G). We list products of congruences in the next theorem.

Theorem 1.5.

(i) $H_\alpha H_\beta = H_{\alpha\beta}$.

(ii) $T_g T_h = T_{gh}$.

(iii) $R^2 = I^2 = \text{identity}$.

(iv) $H_\alpha R = RH_\alpha$.

(v) $H_\alpha T_g = T_{\alpha(g)}H_\alpha$.

(vi) $T_g R = RT_{g^{-1}}H_\alpha$, where α is the inner automorphism $\alpha(x) = gxg^{-1}$.

(vii) $H_\alpha I = IH_\alpha$.

(viii) $(IR)^3 = $ identity.

(ix) $T_g I[\theta] = IT_{\theta^{-1}(g)}[\theta]$

Proof. Routine calculations. ∎

Groups of congruences can be used to classify orthomorphisms and orthomorphism graphs. Some examples of this will be seen in later chapters. In particular, in Chapter 3 we will define and study the orthomorphism graphs $\mathfrak{C}_e \subseteq \text{Orth}(\text{GF}(q)^+)$. If $e \mid q - 1$ then let H be the unique subgroup of $\text{GF}(q)^*$ of order e and define $\mathbf{H}_e = \{H_\alpha : \alpha(x) = ax, a \in H\}$. Then an orthomorphism θ of $\text{GF}(q)^+$ is in $\mathfrak{C}_e$ if and only if $H_\alpha[\theta] = \theta$ for all $H_\alpha \in \mathbf{H}_e$. Further, in chapter 4 we will study the orthomorphism graph $\mathcal{C}(G)$, consisting of those orthomorphisms of G that are also automorphisms of G. These are precisely the orthomorphisms of G for which $T_g[\theta] = \theta$ for all $g \in G$. In section 4 of this chapter we will show how congruences have been used to classify neofields.

In studying orthomorphism graphs, what type of questions should we be asking? Given a group G and an orthomorphism graph $\mathcal{H}$ of G, the problems that concern us are as follows.

i)   Determine exact values of, or bounds for, $\omega(\mathcal{H})$.

ii)   What can we say about the structure of $\mathcal{H}$?

iii)   Can $\mathcal{H}$ contain a complete set of orthomorphisms?

iv)   Given a clique of $\mathcal{H}$ can we extend it to a larger clique of $\mathcal{H}$ or $\text{Orth}(G)$?

While we are interested in the special case $\mathcal{H} = \text{Orth}(G)$, we have few tools available for treating this case. We have data only for small groups, which will be presented in Chapter 6, and this data is mainly the result of computer searches. Thus this case seems for the moment to be beyond us. We therefore restrict ourselves to finding interesting orthomorphism graphs for which these problems might be tractable.

Section 2.   Latin squares and difference matrices.

In this section we will describe applications of cliques of orthomorphism graphs in the construction of mutually orthogonal Latin squares and maximal sets of mutually orthogonal Latin squares. We shall also describe the close relationship that exists between difference matrices and cliques of orthomorphism graphs, and between generalized Hadamard matrices and complete sets of orthomorphisms.

While the study of orthomorphism graphs and their structures is of intrinsic interest, the original impetus for their study arose from the desire to construct large sets of mutually orthogonal Latin squares. A *Latin square of order n* is an $n \times n$ matrix with entries from a symbol set of order n such that each symbol appears exactly once in each row and exactly once in each column.

Example 1.3. Let $g_1, \ldots, g_n$ be the elements of a group of order n. Then the matrix with ijth. entry $g_i g_j$ is a Latin square, which we will refer to as the *Cayley table* of the group.

We say that two Latin squares L_1 and L_2, of the same order and on the same symbol set, are orthogonal if for any pair of symbols a, b there is a uniquely determined pair of integers i, j such that the ijth. entry of L_1 is a while the ijth. entry of L_2 is b. Essentially, if the two Latin squares are superimposed then each ordered pair of symbols will appear exactly once.

Example 1.4. Suppose that F is the finite field of order q. Let $0 = f_1, \ldots, f_q$ be the elements of F. For $k = 2, \ldots, q$ define a matrix L_k whose ijth. entry is $f_k f_j + f_i$. Then $L_2, \ldots, L_q$ form a set of q - 1 mutually orthogonal Latin squares of order q.

Note that each of these Latin squares is obtained from the Cayley table of F^+ by permuting columns. A natural correspondence exists between cliques of orthomorphism graphs and sets of mutually orthogonal Latin squares obtained from Cayley tables by permuting columns.

Let L be the Cayley table of G, $|G| = n$, and let θ be a mapping $G \to G$. Define $L(\theta)$ to be the $n \times n$ matrix with ijth entry $g_i \theta(g_j)$. Then it is easy to establish that $L(\theta)$ is a Latin square if and only if θ is a permutation of G, that if θ is a permutation of G then $L(\theta)$ can be obtained from L by permuting columns, and that if M can be obtained from L by permuting columns then $M = L(\theta)$ for some permutation θ of G.

Lemma 1.1. Let L be the Cayley table of G and let θ, ϕ be permutations of G. Then $L(\theta)$ is orthogonal to L if and only if the mapping $x \to x^{-1}\theta(x)$ is a permutation of G and $L(\theta)$ is orthogonal to $L(\phi)$ if and only if $\theta \sim \phi$.
Proof. Let $g_i g_j = a$ and $g_i \theta(g_j) = b$. Then $g_j^{-1}\theta(g_j) = a^{-1}b$ and j will be uniquely determined if and only if the mapping $x \to x^{-1}\theta(x)$ is a permutation of G. But then i will be uniquely determined also. The proof of the second part is similar. ∎

We shall refer to a set of mutually orthogonal Latin squares obtained from the Cayley table of a group G, by permuting columns, as a set of mutually orthogonal Latin squares *based on G*. The following theorem now needs no proof.

Theorem 1.6. The maximum number of squares possible in a set of mutually orthogonal Latin squares based on G is $\omega(\mathrm{Orth}(G)) + 1$. This maximum is attained.

We shall use N(n) to denote the maximum number of squares possible in a set of mutually orthogonal Latin squares of order n, and R(n) to denote the maximum number of squares possible in a set of mutually orthogonal Latin squares based on a group of order n.

Theorem 1.7.

i)   R(n) = max$\{\omega(\text{Orth}(G)) + 1\colon G$ a group of order $n\}$.

ii)   R(n) $\leq$ N(n) $\leq n - 1$.

Proof.

i)   This follows from Theorem 1.6 and the definition of R(n).

ii)   Given a set of r mutually orthogonal Latin squares of order n, permuting the columns or rows of each square in the same way, changing the symbol set, or permuting the symbols of any given square gives rise to another set of r mutually orthogonal Latin squares of order n. Thus we may assume the symbol set to consist of the integers $\{1, \ldots, n\}$ and the first row of each square to be $1\, 2 \ldots n$. Now the symbol 1 cannot appear in the second row, first column in any square and no other symbol can appear in this position in more than one square and so N(n) $\leq n - 1$. The fact that R(n) $\leq$ N(n) follows immediately from the definitions. $\blacksquare$

We shall call any set of $n - 1$ mutually orthogonal Latin squares of order n a *complete set of mutually orthogonal Latin squares of order n*. For what values of n is R(n) = N(n), and for what values of n is R(n) < N(n)? For n a prime power, Example 1.2 shows R(n) to be $n - 1$ and so R(n) = N(n) = $n - 1$. On the other hand, if n is congruent to 2 modulo 4 then R(n) = 1. This will be proved in Section 5 of this chapter. In this case R(n) < N(n) for $n > 6$ as N(n) ≥ 2 for all $n > 2$ (See for example Dénes and Keedwell (1974, Chapter 11) or Beth, Jungnickel, and Lenz (1984, Chapter IX, Theorem 4.9)).

We shall call a mutually orthogonal set of Latin squares *maximal* if there is no Latin square orthogonal to each square of the set. Maximal sets of orthomorphisms correspond maximal sets of mutually orthogonal Latin squares based on groups. This was proven implicitly in Ostrom (1966).

Theorem 1.8. Let L be the Cayley table of a finite group G. Then $\theta_1, \ldots, \theta_r$ is a maximal clique of Orth(G) if and only if $L, L(\theta_1), \ldots, L(\theta_r)$ is a maximal set of mutually orthogonal Latin squares.

Proof. Let $e = g_1, \ldots, g_n$ be the elements of G. We need only prove that the maximality of $\theta_1, \ldots, \theta_r$ implies the maximality of $L, L(\theta_1), \ldots, L(\theta_r)$. Thus assume $\theta_1, \ldots, \theta_r$ to be a maximal clique of Orth(G). The ijth. entry of L is $g_i g_j$ and the ijth. entry of $L(\theta_k)$ is $g_i\theta_k(g_j)$, $k = 1, \ldots, r$. Assume there exists a Latin square M orthogonal to each of L, $L(\theta_1), \ldots, L(\theta_r)$. We may further assume that the entry of M in row 1 and column 1 is e. Let the $i\phi(i)$th. cells of M be precisely those with entry e and define $\psi(g_i) = (g_{\phi^{-1}(i)})^{-1}$. Then $\theta_1, \ldots, \theta_r, \psi$ is a clique of Orth(G) contradicting the maximality of $\theta_1, \ldots, \theta_r$. $\blacksquare$

Difference matrices and generalized Hadamard matrices are closely related to cliques of orthomorphism graphs. Let $D = \{d_{ij}\}$ be an $r \times \lambda n$ - matrix with entries from a group G of order n. We call D an $(n, r; \lambda, G)$ - *difference matrix* if for each $i, j, i \neq j$, the sequence $\{d_{jk}^{-1}d_{ik}: k = 1, \ldots, \lambda n\}$ contains each element of G exactly λ times. D is a *maximal difference matrix* if there exists no $(n, r + 1; \lambda, G)$ - difference matrix $D' = \{d_{ij}'\}$ satisfying $d_{ij}' = d_{ij}$ for $i = 1, \ldots, r$ and $j = 1, \ldots, \lambda n$. A *generalized Hadamard matrix* is an $(n, \lambda n; \lambda, G)$ - difference matrix. There are various operations that can be performed on difference matrices that always yield difference matrices with the same parameters. We can permute the rows, permute the columns, multiply all the entries, in any given row, on the right by any given element of G, and multiply all the entries, in any given column, on the left by any given element of G. Thus we may assure that every entry in the first row and column of a difference matrix is the identity element of G. There is a natural correspondence between $(n, r; 1, G)$ - difference matrices and $(r - 2)$ - cliques of Orth(G). Let $\theta_1, \ldots, \theta_r$ be a clique of Orth(G) and let $g_1 \ldots, g_n$ be the elements of G. Define an $(r + 2) \times n$ matrix $D = (d_{ij})$ by $d_{1j} = e$ for $j = 1, \ldots, n$, $d_{2j} = g_j$ for $j = 1, \ldots, n$, and $d_{ij} = \theta_{i-2}(g_j)$ for $i = 3, \ldots, r + 2$ and $j = 1, \ldots, n$. Then D is an $(n, r + 2; 1, G)$ - difference matrix. Let D be an $(n, r + 2; 1, G)$ - difference matrix with every entry in the first row and column of D equal to e. Define $\theta_i: G \to G$ by $\theta_i(d_{2j}) = d_{i+2\,j}$ for $i = 1, \ldots, r$. Then $\theta_1, \ldots, \theta_r$ is a clique of Orth(G). It is easy to see that an $(n, r + 2; 1, G)$ - difference matrix is maximal if and only if the corresponding clique of Orth(G) is maximal. For more information on difference matrices and generalized Hadamard matrices see the survey papers by De Launey (1986, and 1987), and Jungnickel (1979), the book by Beth, Jungnickel, and Lenz (1985, Chapter VIII), and the papers reprinted in Evans (To appear - a).

Example 1.5. Jungnickel (1980) constructed $(p^n, r + 1; 1, Z_{pn})$ - difference matrices, p a prime, $1 \le r \le p - 1$, by setting the ijth entry equal to $(i - 1)(j - 1)$ modulo p^n, for $i = 1, \ldots, r + 1$ and $j = 1, \ldots, p^n$. The corresponding $(r - 2)$ - clique of Orth(Z_{pn}), is $\{x \to mx: m = 2, \ldots, r\}$. For $r = p - 1$ this difference matrix is maximal. Thus there exists a maximal set of $p - 1$ mutually orthogonal Latin squares of order p^n. This will be proved in Chapter 5, Section 4.

Theorem 1.9. Jungnickel (1979). Let G be an abelian group. Then H is a generalized Hadamard matrix over G if and only if its transpose is also a generalized Hadamard matrix over G.
Proof. See Jungnickel (1979, Theorem 2.2) or Brock (1988, Theorem 4.1). ∎

Theorem 1.10. Suppose that a group G, of order nm, admits a homomorphism ϕ onto a group H, of order m, and let $D = \{d_{ij}\}$ be an $(nm, r; 1, G)$ - *difference matrix*. Then the matrix $D' = \{\phi(d_{ij})\}$ is an $(nm, r; n, H)$ - *difference matrix*. In particular if G admits a complete set of orthomorphisms then there exists a generalized Hadamard matrix of order mn over H.
Proof. Routine calculation. ∎

An immediate corollary.

Corollary 1.4. Let $q = p^r$, p a prime. Then there exists a generalized Hadamard matrix of order $p^s q$ over $GF(q)^+$, for all non-negative integers s.

Theorem 1.10 is not of great use in the construction of generalized Hadamard matrices, as the only groups that are known to admit complete sets of orthomorphisms are the elementary abelian groups. It is, however, useful for establishing the non-existence of complete sets of orthomorphisms for many groups. We shall see examples of this in Chapter 5, Section 5.

Section 3. Nets and affine planes.

The well known correspondence between mutually orthogonal Latin squares and nets or equivalently transversal designs, and between complete sets of Latin squares and affine (equivalently projective) planes, implies that orthomorphisms can be used in the construction of such incidence structures. We will give direct constructions of these incidence structures from cliques of orthomorphism graphs.

A *net* of order n and degree k is an incidence structure with n^2 points and nk lines, partitioned into k parallel classes of n lines each, satisfying the following properties.
i)   Each line is incident with n points.
ii)   Each point is incident with k lines.
iii)   Two distinct lines have no points in common if they are in the same parallel class and exactly one point in common if they are in different parallel classes.

Consider the dual of a net of order n and degree k. This is an incidence structure with nk lines and n^2 points, partitioned into k point classes of n points each, each point incident with n lines, each line incident with k points, two distinct points being joined by a unique line if in different point classes and no line if in the same point class. Such an incidence structure is called a *transversal design* of order n and degree k. A transversal design is *resolvable* if its lines can be partitioned into parallel classes, the lines of each parallel class partitioning the point set.

A net of order n and degree $n + 1$ is called an *affine plane* of order n. A *projective plane* of order n is an incidence structure with $n^2 + n + 1$ points and $n^2 + n + 1$ lines, $n + 1$ points on each line, and $n + 1$ lines through each point, each pair of points being incident with a unique common line, and each pair of lines being incident with a unique common point. It is well known that if we remove a line, and all points incident with it, from a projective plane of order n the incidence structure obtained is an affine plane of order n. By reversing this construction we can construct a projective plane of order n from an affine plane of order n as follows. For each parallel class P adjoin an ideal point $[P]$, incident with each line of P but no line of any other parallel class. Adjoin a new line, the ideal line, incident with each of the ideal points. The ideal points are often referred to as points at infinity and the ideal line as the line at infinity.

For more information on nets, transversal designs, affine and projective planes see Dembowski (1968), Dénes and Keedwell (1974), and Beth, Jungnickel, and Lenz (1985).

Example 1.6. Let F be a finite field of order q and define an incidence structure as follows. The points will be ordered pairs of elements of F and the lines will be described by equations $x = c$, and $y = mx + b$, where m, b, and $c \in F$. This is the *Desarguesian* affine plane of order q. The parallel classes are $\{x = c: c \in F\}$, and $\{y = mx + b: b \in F\}$, $m \in F$. If $r < q + 1$ then a net of order q and degree r can be constructed from this affine plane by removing all lines from $q + 1 - r$ parallel classes.

We have thus shown that there exist affine planes of order q, whenever q is a prime power. No planes of non-prime power order are known and it is conjectured that none exist. For planes of prime power order a great many non-Desarguesian planes have been constructed. However, the only known planes of prime order are Desarguesian and it is conjectured that no non-Desarguesian planes of prime order exist.

Let G be a finite group, which we shall write additively with identity 0, whether abelian or not. Let 1 be a given element of $G - \{0\}$, let S be a subset of the elements of G, where $0, 1 \in S$, and let $\mathbb{F} = \{\phi_x: x \in S\}$, where ϕ_0 is defined by $\phi_0(x) = 0$ for all x, ϕ_1 is defined by $\phi_1(x) = x$ for all x, and $\phi_x: G \to G$ is a mapping satisfying $\phi_x(1) = x$ and $\phi_x(0) = 0$. We may think of $\phi_m(x)$ as defining a product $x \cdot m$ of elements of G. Define an incidence structure $N(\mathbb{F})$ as follows. The point set of $N(\mathbb{F})$ is $\{(x, y): x, y \in G\}$. The lines are $x = a$ and $y = \phi_m(x) + b$. Define a second incidence structure $T(\mathbb{F})$ in a similar manner. The points of $T(\mathbb{F})$ are the ordered pairs (i, b), $i \in S$, $b \in G$, and the lines are the sets B_{xy}, $x, y \in G$, $B_{xy} = \{(i, y + \phi_i(x)): i \in S\}$, incidence being set inclusion. $T(\mathbb{F})$ is the dual of the incidence structure obtained from $N(\mathbb{F})$ by removing the lines $x = a$. To see this equate the point $(x, -y)$ of $N(\mathbb{F})$ with the line B_{xy} of $T(\mathbb{F})$ and the line $y = \phi_i(x) - b$ of $N(\mathbb{F})$ with the point (i, b) of $T(\mathbb{F})$.

Theorem 1.11. $N(\mathbb{F})$ is a net if and only if $\{\phi_x: x \in S - \{0, 1\}\}$ is a clique of $\mathrm{Orth}(G)$, in which case $N(\mathbb{F})$ will have order $|G|$ and degree $|S| + 1$.
Proof. Let $k = |S| + 1$.

Let $P_i = \{y = \phi_i(x) + b: b \in G\}$ and $P_\infty = \{x = a: a \in G\}$. For convenience we shall call these parallel classes.

Note that $N(\mathbb{F})$ contains n^2 points and nk lines, the lines being partitioned into k parallel classes of n lines each. Each line passes through exactly n points. The lines of each parallel class partition the point set of $N(\mathbb{F})$ and each line not in P_∞ intersects each line in P_∞ in exactly one point.

Thus $N(\mathbb{F})$ is a net if and only if lines from distinct parallel classes P_i, $i = 0, \dots, k - 1$, intersect in exactly one point.

For $i \neq 0$, a line $y = \phi_i(x) + b$ intersects each line of P_0 in exactly one point if and only if ϕ_i is a permutation.

For $i \neq 0, 1$, a line $y = \phi_i(x) + b$ intersects each line $y = x + c$ in exactly one point if and only if for each c the equation $-x + \phi_i(x) = c - b$ uniquely determines x, i.e. if and only if $\phi_i \in \text{Orth}(G)$.

For $i, j \neq 0, 1, i \neq j$, each line of P_i intersects each line of P_j in exactly one point if and only if for each b the equation $-\phi_i(x) + \phi_j(x) = b$ uniquely determines x, i.e. if and only if $\phi_i \sim \phi_j$. Hence the result. ∎

In the course of proving Theorem 1.11 we have established the following.

Corollary 1.5. i) Repphun (1965). $N(\mathbb{F})$ is an affine plane if and only if $S = G$ and $\{\phi_x : x \in S - \{0, 1\}\}$ is a complete set of orthomorphisms of G.

ii) Jungnickel (1979). $T(\mathbb{F})$ is a resolvable transversal design if and only if $\{\phi_x : x \in S - \{0, 1\}\}$ is a clique of $\text{Orth}(G)$, in which case $T(\mathbb{F})$ will have order $|G|$ and degree $|S|$.

Corollary 1.6. Let G be a group of order n. G cannot admit a complete set of orthomorphisms if $n \equiv 1$ or 2 modulo 4 and n is not the sum of two squares.
Proof. This follows from Corollary 1.5 and the Bruck-Ryser theorem (See Dembowski (1968, p. 144)). ∎

A *collineation* of a net, transversal design, or affine plane is any permutation of the points that induces a permutation of the lines that preserves incidence. Nets and affine planes constructed from orthomorphisms admit a special class of collineations. Specifically the mapping $(x, y) \rightarrow (x, y + g)$ is a collineation for all $g \in G$. The induced line permutations are $x = c \rightarrow x = c$, and $y = \phi_m(x) + b \rightarrow y = \phi_m(x) + (b + g)$. These collineations form a group under composition. Each collineation of this group fixes any line of the form $x = \text{constant}$, fixes all parallel classes, and the group acts sharply transitively on the points of any line of the form $x = \text{constant}$. Affine planes that admit such a collineation group we shall call *Cartesian planes* and the corresponding collineation group we shall call a *Cartesian group*. The existence of such a collineation group actually characterizes nets and affine planes constructed from orthomorphisms. A collineation of an affine plane that fixes all parallel classes, and all lines of one parallel class is called a *translation*, and the parallel class whose lines are fixed by the translation is called the *direction* of the translation.

Theorem 1.12. a) Let N be a net of degree $r + 3$ and let G be a group of collineations of N that fixes all lines of one parallel class, acts sharply transitively on the points of any line in this class, and fixes all parallel classes. Then there exists an r-clique of $\text{Orth}(G)$ such that the net constructed from this clique is isomorphic to N.

b) Let T be a resolvable transversal design of degree $r + 2$ and let G be a group of collineations of T that fixes all point classes, acts sharply transitively on the points of any point class, and semiregularly on the lines of T. Then there exists an r-clique of $\text{Orth}(G)$ such that the transversal design constructed from this clique is isomorphic to T.
Proof. See Jungnickel (1979). ∎

Corollary 1.7. Repphun (1965). Let A be a Cartesian plane and let G be the corresponding Cartesian group. Then $A = N(\mathbb{F})$ for some $\mathbb{F}$, where $\mathbb{F} - \{\phi_0, \phi_1\}$ is a complete set of orthomorphisms of G.

As we know certain automorphisms of $\text{Orth}(G)$ we might ask what effect these have on the net $N(\mathbb{F})$. This will give us some insight into the relationship between automorphisms of $\text{Orth}(G)$ and collineations of $N(\mathbb{F})$.

Theorem 1.13. Let T be a translation, homology, or reflection and let N be the net constructed from the clique $\phi_2, \dots, \phi_r$ of $\text{Orth}(G)$. Then the net N' constructed from $T[\phi_2]$, $\dots, T[\phi_r]$ is isomorphic to N.

Proof. If $T = T_g$ then it is routine to check that the following is an isomorphism.

$(x, y) \rightarrow (x', y') = (x - g, y)$

$x = c \rightarrow x' = c - g$

$y = \phi_m(x) + b \rightarrow y' = T_g[\phi_m](x') + (\phi_m(g) + b).$

   If $T = H_\alpha$ then again it is routine to check that the following is an isomorphism.

$(x, y) \rightarrow (x', y') = (\alpha(x), \alpha(y))$

$x = c \rightarrow x' = \alpha(c)$

$y = \phi_m(x) + b \rightarrow y' = H_\alpha[\phi_m](x') + \alpha(b).$

   The case $T = R$ is similar and the corresponding isomorphism is given below.

$(x, y) \rightarrow (x', y') = (-x, -x + y)$

$x = c \rightarrow x' = -c$

$y = \phi_m(x) + b \rightarrow y' = R[\phi_m](x') + b.$ ∎

From the isomorphisms described in the proof to Theorem 1.13 we can read off information about collineations of nets constructed from orthomorphisms. Some of this information is presented in the following corollary.

Corollary 1.8. Let N be the net constructed from the clique $\phi_2, \dots, \phi_r$ of $\text{Orth}(G)$.

i)   If $T_g[\phi_m] = \phi_m$ for all m then the mapping $(x, y) \rightarrow (x - g, y)$ is a collineation of N.

ii)   If $H_\alpha[\phi_m] = \phi_m$ for all m then the mapping $(x, y) \rightarrow (\alpha(x), \alpha(y))$ is a collineation of N.

iii)   If $R[\phi_m] = \phi_m$ for all m then the mapping $(x, y) \rightarrow (-x, -x + y)$ is a collineation of N.

The last of these collineations applies to few nets due to the next result.

Theorem 1.14. There exists an r-clique of $\text{Orth}(G)$ in which each orthomorphism is fixed by R if and only if $|G|$ is odd and $r = 1$.

Proof. Let θ be an orthomorphism fixed by R. Then $x^{-1}\theta(x) = \theta(x^{-1})$ and so $x = x^{-1}$ if and only if $x = e$. Thus $|G|$ must be odd, and if $n = |G|$ is odd then the mapping $x \rightarrow x^{(n+1)/2}$ is an orthomorphism of G, fixed by R.

Let θ and ϕ be two orthomorphisms fixed by R. These cannot be adjacent as $\theta(x^{-1})^{-1}\phi(x^{-1}) = (x^{-1}\theta(x))^{-1}(x^{-1}\phi(x)) = \theta(x)^{-1}\phi(x)$. $\blacksquare$

If $S = G$ the the dual of $\mathbb{F}$ is $\mathbb{F}^*$ where $\phi^*_x \in \mathbb{F}^*$ is defined by $\phi^*_x(y) = \phi_y(x)$.

Theorem 1.15. $N(\mathbb{F}^*)$ and $N(\mathbb{F})$ are both affine planes if and only if the elements of $\mathbb{F} - \{\phi_0, \phi_1\}$ form a complete set of orthomorphisms of G, and the mapping $x \rightarrow \phi_y\phi_z^{-1}(x) - x$ is a permutation whenever $y, z \neq 0, 1, y \neq z$.

Proof. It is easy to show that $\phi^*_0 = \phi_0, \phi^*_1 = \phi_1, \phi^*_x(0) = 0$, and $\phi^*_x(1) = x$. Let $\phi^*_a, \phi^*_b \in \mathbb{F}^*$. Then $- \phi^*_a(c) + \phi^*_b(c) = - \phi^*_a(d) + \phi^*_b(d)$ if and only if $- \phi_c(a) + \phi_c(b) = - \phi_d(a) + \phi_d(b)$ if and only if $\phi_d(a) - \phi_c(a) = \phi_d(b) - \phi_c(b)$ if and only if $a = b$ or $c = d$. For $c, d \neq 0$, setting $x = \phi_c(a)$ and $y = \phi_c(b)$ we see that $\phi_d(a) - \phi_c(a) = \phi_d(b) - \phi_c(b)$ if and only if $\phi_d\phi_c^{-1}(x) - x = \phi_d\phi_c^{-1}(y) - y$. Hence the result. $\blacksquare$

Corollary 1.9. If G is abelian then $N(\mathbb{F})$ is an affine plane if and only if $N(\mathbb{F}^*)$ is also an affine plane.

Note that Corollary 1.9 is actually a special case of Theorem 1.9. Finally let us mention some orthomorphism characterizations of certain types of affine planes. As orthomorphisms have not actually been used in the study of these classes of planes we refer the interested reader to Dembowski (1968) for their definitions. Each of these planes can be coordinatized by quasifields and can be characterized by the properties of the corresponding quasifields. Equating the product $x \cdot m$ with $\phi_m(x)$, the orthomorphism characterizations then follow immediately from the quasifield characterizations.

Theorem 1.16. Let $\mathbb{F} = \{\phi_x: x \in G - \{0, 1\}\} \cup \{\phi_0, \phi_1\}$, where $\mathbb{F} - \{\phi_0, \phi_1\}$ is a complete set of orthomorphisms of G.
(i) $N(\mathbb{F})$ is a translation plane if and only if $\phi_x \in \text{Aut}(G)$ for all $x \neq 0$.
(ii) $N(\mathbb{F})$ is a dual translation plane if and only if $(\mathbb{F}, +)$ is a group, in which case $\phi_x + \phi_y = \phi_{x+y}$ and $(\mathbb{F}, +) \cong G$.
(iii) $N(\mathbb{F})$ is a nearfield plane if and only if $\phi_x \in \text{Aut}(G)$ for all $x \neq 0$ and $\mathbb{F} - \{\phi_0\}$ is a group under the operation of composition.
(iv) $N(\mathbb{F})$ is a semifield plane if and only if $(\mathbb{F}, +)$ is a group, in which case $\phi_x + \phi_y = \phi_{x+y}$ and $(\mathbb{F}, +) \cong G$, and $\phi_x \in \text{Aut}(G)$ for $x \neq 0$.
(v) $N(\mathbb{F})$ is a Desarguesian affine plane if and only if G, with multiplication defined by $xm = \phi_m(x)$, is a field.

A proof of Theorem 1.16 can be found in Repphun (1965) and proofs of parts i) and ii) can also be found in Evans (1989d).

Section 4. Neofields.

In this section we will define (left) neofields and near orthomorphisms, a generalization of the concept of orthomorphism. It will be shown that near orthomorphisms are equivalent to left neofields and as a consequence that orthomorphisms are equivalent to left neofields in which $1 + 1 = 0$. We will establish a strong relationship between properties of left neofields and properties of the corresponding near orthomorphisms. In fact, we will find that the properties of a left neofield, in which $1 + 1 = 0$, are determined by the congruences of Orth(G) that fix the corresponding orthomorphism. Similar results will be proved for left neofields, in which $1 + 1 \neq 0$, and their corresponding near orthomorphisms.

A *near orthomorphism* of a group G is a bijection θ: $G - \{t\} \to G - \{e\}$, for which the mapping $x \to x^{-1}\theta(x)$ is also a bijection from $G - \{t\}$ onto $G - \{e\}$. We call t the *exdomain element* of θ. Orthomorphisms can be regarded as a special class of near orthomorphisms as if the exdomain element of the near orthomorphism θ is e then θ becomes an orthomorphism when we define $\theta(e) = e$. The definition given here differs subtly from that of Hsu (1991). Hsu defines a near orthomorphism of a group G to be a bijection θ: $G \to G$ for which the set $\{x^{-1}\theta(x): x \in G\} = G - \{e\}$. If $t \neq e$ is the exdomain element of a near orthomorphism θ then setting $\theta(t) = e$ yields a mapping that satisfies Hsu's definition of a near orthomorphism. It should be noted that our definition is consistent with the definition of a (K, λ) - near orthomorphism introduced in Hsu and Keedwell (1984). This is actually called a (K, λ) - near complete mapping in their paper, but both coauthors have since decided to use the term orthomorphism instead. It was proved implicitly in the work of Paige (1947b) that, if G is abelian then the exdomain element of a near orthomorphism of G must be the identity if the Sylow 2-subgroup of G is trivial or noncyclic, or the unique element of order 2 if the Sylow 2-subgroup is nontrivial and cyclic.

A *left neofield* is a set N with two binary operations, addition and multiplication, satisfying the following:

i)   The elements of N form a loop under addition, with identity 0.
ii)   The nonzero elements of N form a group under multiplication, with identity 1.
iii) $a(b + c) = ab + ac$ for all $a, b, c \in N$.

A *left neofield* is called a *neofield* if the right distributive law is also satisfied. Clearly if the multiplicative group is abelian then the terms neofield and left neofield are synonymous. The *order* of the left neofield N, denoted $|N|$, is the number of elements of N, i.e. one plus the order of its multiplicative group. A left neofield is completely determined by its multiplicative group and the mapping $\theta(x) = 1 + x$, called the *presentation function* of the left neofield.

Neofields were first introduced by Paige (1949). Bruck (see Paige (1949), Theorem I.1) implicitly established the connection between neofields with multiplicative group G and near orthomorphisms of G fixed by the homologies H_α, $\alpha \in \text{Inn}(G)$. Later Hsu and Keedwell (1984) generalized this result to show that there is a 1-1 correspondence between left neofields with multiplicative group G and near orthomorphisms of G.

Theorem 1.17. Hsu and Keedwell (1984). Let G be a group, written multiplicatively with identity 1. Let θ be a near orthomorphism of G, with exdomain element t, and extend θ to a bijection $G \cup \{0\} \to G \cup \{0\}$ by setting $\theta(t) = 0$ and $\theta(0) = 1$. Then θ is the presentation function of a left neofield.

Conversely, Let θ be the presentation function of a left neofield with multiplicative group G. Let t be the unique solution to the equation $\theta(x) = 0$. Then θ restricted to $G - \{t\}$ is a near orthomorphism of G with exdomain element t.

Proof. Let θ be a near orthomorphism of a group G, with exdomain element t, and extend θ to a bijection $G \cup \{0\} \to G \cup \{0\}$ by setting $\theta(t) = 0$ and $\theta(0) = 1$. Let $N = G \cup \{0\}$ and define addition and multiplication in N as follows. Multiplication is as in G except that $0a = a0 = 0$ for all $a \in N$. To define addition, $x + y = y$ if $x = 0$, and $x\theta(x^{-1}y)$ if $x \neq 0$.

We see that $0 + y = y$ for all $y \in N$ and if $y \neq 0$ then $y + 0 = y\theta(0) = y$. Suppose that $a + b = c$. Given a and b the value of c is uniquely determined. Given a and c then $b = c$ if and only if $a = 0$, and b is uniquely determined by the equation $\theta(a^{-1}b) = a^{-1}c$ if $a \neq 0$. Given b and c then $a = 0$ if and only if $b = c$. Otherwise a is uniquely determined by the equation $(a^{-1}b)^{-1}\theta(a^{-1}b) = b^{-1}c$. Thus N is a loop under addition. Next consider $a(b + c)$. This must equal $ab + ac$ if any of a, b, or c is 0. If $a, b, c \neq 0$ then

$$a(b + c) = (ab)\theta((ab)^{-1}ac) = ab + ac$$ and so the left distributive law holds. Thus N is a left neofield.

Conversely, if θ is the presentation function of a left neofield N with multiplicative group G then $\theta(t) = 0$ for a unique element t of N. It is easily seen that θ is a bijection from $G - \{t\}$ onto $G - \{1\}$ and the mapping $x \to x^{-1}\theta(x) = x^{-1} + 1$ is a bijection from $G - \{t\}$ onto $G - \{1\}$. Thus the restriction of θ to $G - \{t\}$ is a near orthomorphism of G. ∎

Corollary 1.10. There is a 1-1 correspondence between orthomorphisms of a group G and left neofields with multiplicative group G in which $1 + 1 = 0$.

We remark that for a left neofield N the mapping $a \to ga$ is an automorphism of the additive loop of N, for each nonzero element g of N. In fact, a loop can be the additive loop of a left neofield if and only if it admits an automorphism group that acts sharply transitively on its nonidentity elements.

Let us define an *automorphism* of a left neofield N to be a bijection $\alpha: N \to N$ for which $\alpha(a + b) = \alpha(a) + \alpha(b)$, and $\alpha(ab) = \alpha(a)\alpha(b)$, for all $a, b \in N$. Clearly the automorphism group of a left neofield is a subgroup of the automorphism group of its multiplicative group, as well as a subgroup of the automorphism group of its additive loop.

We shall define homologies for near orthomorphisms in the same way as for orthomorphisms, by setting $H_\alpha[\theta] = \alpha\theta\alpha^{-1}$, $\alpha \in \text{Aut}(G)$. As before $H_\alpha H_\beta = H_{\alpha\beta}$.

Theorem 1.18. Let θ be a near orthomorphism of G with exdomain element t, let N be the left neofield constructed from θ, and let α be an automorphism of G.

i) $H_\alpha[\theta]$ is a near orthomorphism with exdomain element $\alpha(t)$.

ii) α extends to an automorphism of N, by setting $\alpha(0) = 0$, if and only if $H_\alpha[\theta] = \theta$.

iii)   If $\alpha(x) = c^{-1}xc$ then $H_\alpha[\theta] = \theta$ if and only if $(a + b)c = ac + bc$ for all $a, b \in N$.

iv) N is a neofield if and only if $H_\alpha[\theta] = \theta$ for all $\alpha \in \text{Inn}(G)$.

Proof.

i)   This follows as $x^{-1}\alpha\theta\alpha^{-1}(x) = \alpha(\alpha^{-1}(x)^{-1}\theta\alpha^{-1}(x))$.

ii)   If $a \neq 0$ then $\alpha(a + b) = \alpha(a) + \alpha(b)$ if and only if $\alpha(a\theta(a^{-1}b)) =$
$\alpha(a)\theta(\alpha(a^{-1}b))$. By setting $x = \alpha(a^{-1}b)$, this is seen to be true if and only if
$\alpha\theta\alpha^{-1}(x) = \theta(x)$. Hence the result.

iii)   If any of $a, b,$ or c is zero then $(a + b)c = ac + bc$. If $a, b, c \neq 0$ then $(a + b)c$
$= a\theta(a^{-1}b)c$ and $ac + bc = ac\theta(c^{-1}a^{-1}bc)$ and $a\theta(a^{-1}b)c = ac\theta(c^{-1}a^{-1}bc)$ if and
only if $H_\alpha[\theta] = \theta$.

iv)   This is an immediate consequence of ii). ∎

Corollary 1.11. If θ is a near orthomorphism of G corresponding to a neofield then the exdomain element of θ is in the center of G.

We can extend R and I to the domain of near orthomorphisms as follows. If θ is a near orthomorphism then $R[\theta]$: $x \to x\theta(x^{-1})$ and $I[\theta]$: $x \to t^{-1}\theta^{-1}(tx)$, where t is the exdomain element of θ, are near orthomorphisms with exdomain element t^{-1}. The relationships $R^2 = $ identity, $H_\alpha R = RH_\alpha$, $I^2 = $ identity, and $H_\alpha I = IH_\alpha$ still hold. The relationship $(IR)^3 = $ identity holds if $t \in Z(G)$ and $t^2 = 1$.

For a left neofield N we define the following properties.

i) N is *commutative* if addition is commutative.

ii) N has the *right inverse property* if for all $a \in N$ there exists $(-a)_R \in N$ such that
$(x + a) + (-a)_R = x$ for all $x \in N$.

iii) N has the *left inverse property* if for all $a \in N$ there exists $(-a)_L \in N$ such that
$(-a)_L + (a + x) = x$ for all $x \in N$.

iv) N has the *inverse property* if it has both the left and right inverse properties.

v) N has the *exchange inverse property* if for all $a \in N$ there exists $(-a)_L \in N$ such that
$(-a)_L + (x + a) = x$ for all $x \in N$.

Correspondences between properties of left neofields and mappings that fix the corresponding near orthomorphisms are established in the next theorem.

Theorem 1.19. Let θ be a near orthomorphism of G with exdomain element t, and let N be the left neofield constructed from θ.

i) N is commutative if and only if $R[\theta] = \theta$.

ii) If $t \in Z(G)$ then N has the right inverse property if and only if $IRI[\theta] = \theta$.

iii) N has the left inverse property if and only if $I[\theta] = \theta$.

iv) If $t \in Z(G)$ then N has the inverse property if and only if $I[\theta] = \theta$ and $IRI[\theta] = \theta$.

v) N has the exchange inverse property if and only if $RI[\theta] = \theta$.

Proof. Let t be the exdomain element of θ. Thus $1 + t = 0$. We will first establish that a necessary condition for N to be commutative or have the left or right inverse property is that $t^2 = 1$. Note that $t + 1 = 0$ if and only if $t(1 + t^{-1}) = 0$ if and only if $t^{-1} = t$. Thus N commutative implies that $t + 1 = 1 + t = 0$ and so $t^2 = 1$. If N has the right inverse property then $(-t)_R = 1$ as $(1 + t) + (-t)_R = 1$ and then $(0 + t) + 1 = 0$ again implies that $t^2 = 1$. If N has the left inverse property then $(-1)_L = t$ as $(-1)_L + (1 + t) = t$ and then $t + (1 + 0) = 0$ again implies that $t^2 = 1$.

It is also important to note that $T[\theta] = \theta$, for T = I, R, and IRI, only if $t^2 = 1$. Thus, in the proofs of i) , ... , iv), we shall assume that $t^2 = 1$.

i) If either a or $b = 0$, or $a = bt$, or $b = at$, then $a + b = b + a$. Otherwise $a + b = b + a$ if and only if $a\theta(a^{-1}b) = b\theta(b^{-1}a)$ if and only if $\theta(a^{-1}b) = (a^{-1}b)\theta((a^{-1}b)^{-1})$ if and only if $R[\theta](a^{-1}b) = \theta(a^{-1}b)$.

ii) If N has the right inverse property then $(-a)_R = at$. If $a = 0$, $x = 0$, or $x = at$, then $(x + a) + at = x$. Otherwise $(x + a) + at = x$ if and only if $\theta(\theta(x^{-1}a)^{-1}x^{-1}at) = \theta(x^{-1}a)^{-1}$ if and only if $\theta(y\theta^{-1}(ty^{-1})) = ty$, where $y = \theta(x^{-1}a)^{-1}t$, if and only if $ty\theta^{-1}(ty^{-1}) = t\theta^{-1}(ty)$ if and only if $RI[\theta](y) = I[\theta](y)$ if and only if $IRI[\theta] = \theta$.

iii) If N has the left inverse property then $(-a)_L = at$. If $a = 0$, $x = 0$, or $x = at$, then $at + (a + x) = x$. Otherwise $at + (a + x) = x$ if and only if $t\theta(t\theta(a^{-1}x)) = a^{-1}x$ if and only if $I[\theta](a^{-1}x) = \theta(a^{-1}x)$.

iv) This follows from ii) and iii).

v) If N has the exchange inverse property then $(-a)_L = at^{-1}$. If $a = 0$, $x = 0$, or $x = at^{-1}$, then $at^{-1} + (x + a) = x$. Otherwise $at^{-1} + (x + a) = x$ if and only if $at^{-1}\theta(ta^{-1}x\theta(x^{-1}a)) = x$ if and only if $RI[\theta](a^{-1}x) = \theta(a^{-1}x)$. ∎

Note that, by Theorem 1.14, if $1 + 1 = 0$ in a commutative left neofield then the order of the left neofield must be even. Further correspondences between the properties of neofields and properties of the corresponding near orthomorphisms can be found in Keedwell (1983).

These properties were used by Hsu (1980) to classify cyclic neofields, i.e. neofields in which the multiplicative group is cyclic. In a cyclic neofield $t = 1$ if the neofield is of even order, or the unique element of multiplicative order 2 if the order of the neofield is odd. Thus the group <I, R> is formally isomorphic to the dihedral group of order 6. This group has six

subgroups, which gives rise to a classification of cyclic neofields into six different types.
CIP-neofields are commutative and have the inverse property, which is equivalent to saying that their corresponding near orthomorphisms are fixed by each element of <I, R>.
XIP-neofields have the exchange inverse property: Their corresponding near orthomorphisms are fixed by RI. LXP-neofields have the left inverse property,: Their corresponding near orthomorphisms are fixed by I. CMP-neofields are commutative: Their corresponding near orthomorphisms are fixed by R. RXP-neofields have the right inverse property: Their corresponding near orthomorphisms are fixed by IRI. Any cyclic neofield is an
XMP-neofield. Ordered by inclusion, these classes form a lattice that is isomorphic to the subgroup lattice of <I, R>. Hsu (1980) gives constructions for each such type of cyclic neofield.

Section 5. The existence problem.

The problem presents itself as to which groups admit orthomorphisms or equivalently complete mappings. For infinite groups, this problem was settled by Bateman (1950). He proved that all infinite groups admit complete mappings. For finite groups this question is still open. Hall and Paige (1955) proved that a finite group with a non-trivial, cyclic Sylow 2-subgroup cannot admit orthomorphisms. To prove their result we first need a lemma, due to Burnside.

Lemma 1.2. If the Sylow 2-subgroup of a finite group is non-trivial and cyclic then the group admits a homomorphism onto its Sylow 2-subgroup.
Proof. See Beth, Jungnickel, and Lenz (1985, chapter X, Lemma 12.1) or Huppert (1967, chapter IV, Satz 2.8). ∎

Theorem 1.20. Hall and Paige (1955). A finite group admits no orthomorphisms if its Sylow 2-subgroup is non-trivial and cyclic.
Proof. Let G be a group with a non-trivial, cyclic Sylow 2-subgroup and let $|G| = mn$, n odd, m a power of 2. Then by Lemma 1.2 there exists an epimorphism $\phi: G \rightarrow Z_m$.

If G admits an orthomorphism θ then we have the following.

$$\sum_{g \in G} \phi(g) = \sum_{g \in G} \phi(g^{-1}\theta(g)) = \sum_{g \in G} (\phi\theta(g) - \phi(g)) = \sum_{g \in G} \phi(g) - \sum_{g \in G} \phi(g) = 0$$

But direct computation yields $\displaystyle\sum_{g \in G} \phi(g) \equiv n \sum_{i=0}^{m-1} i \equiv nm(m-1)/2 \not\equiv 0$ (modulo m). A contradiction from which the result follows. ∎

Precursors to the Hall and Paige result were two proofs in 1951; a proof by Bruck (1951, Theorem 7) that if a group G contains a normal subgroup H of odd order, such that the quotient group G/H is a non-trivial cyclic group of even order, then G cannot admit a complete mapping; and a proof by Paige (1951, Theorem 5) that if H is an odd order subgroup of a group G and G/H does not admit complete mappings then G does not admit complete mappings. Either of these results combined with Lemma 1.2 would have yielded an earlier proof of Hall and Paige's theorem.

Some special cases of Hall and Paige's result are as follows.

Corollary 1.12. Euler (1779). A cyclic Latin square of even order has no transversals and hence no orthogonal mate.

Proof. This is equivalent to proving that a cyclic group of even order cannot admit orthomorphisms. ∎

Hurwitz (1882) proved that the multiplicative group of a finite field of odd prime order cannot admit a complete mapping. This is a special case of Corollary 1.12. Euler's result was later rediscovered by Hedayat and Federer (1969).

Another special case.

Corollary 1.13. A group of order $4n + 2$ cannot admit orthomorphisms.

This was first proved by Fleisher (1934), in his thesis. Later Mann (1942 and 1944) gave two different proofs of this result. Another proof of this can be found in Jungnickel (1980).

An unsolved problem: Is the converse to Hall and Paige's theorem true? Hall and Paige conjectured this to be the case and proved their conjecture true for many classes of groups.

Hall-Paige conjecture. A finite group with a trivial or non-cyclic Sylow 2-subgroup admits orthomorphisms.

This conjecture had already been proved for abelian groups.

Theorem 1.21. Paige (1947a and 1947b). A finite abelian group admits orthomorphisms if and only if its Sylow 2-subgroup is trivial or non-cyclic.

Proof. Let G be a finite abelian group with a trivial or non-cyclic Sylow 2-subgroup. Then G admits a subnormal series of the form $\{e\} = G_1 \lhd G_2 \lhd \ldots \lhd G_m = G$, where G_{i+1}/G_i is isomorphic to either an elementary abelian 2-group of order at least 4, a group of odd order, or a group of the form $Z_2 \times Z_m$, $m = 2^n$, $n > 1$. For abelian groups it is easy to show that if H is a subgroup of G and H and G/H both admit orthomorphisms then so does G. (This will be proved for all groups in Corollary 1.15.) Thus to prove existence we need only show that $Z_2 \times Z_m$, $m = 2^n$, $n > 1$, admits orthomorphisms. Let $Z_2 = <h>$ and $Z_m = <g>$. Then the following table defines an orthomorphism of $Z_2 \times Z_m$.

x	$\rightarrow$	$\theta(x)$	
hg^i	$\rightarrow$	g^{2i}	$i \in S$
g^i	$\rightarrow$	g^{2i-1}	$i \in S$
hg^i	$\rightarrow$	hg^{2i-1}	$i \notin S$
g^i	$\rightarrow$	hg^{2i}	$i \notin S$

Where $S = \{1, \ldots, 2^{n-1}\}$. The converse follows from Hall and Paige's theorem. ∎

Paige's proof of this theorem was actually non-constructive. Hall (1952) generalized this result by proving that if A is an abelian group of order n and if $b_1, \ldots, b_n \in A$ satisfies $\sum_{i=1}^{n} b_i = 0$ then there exists a labelling $g_1, \ldots, g_n$ of the elements of A and a permutation θ of A such that $\theta(g_i) - g_i = b_i$ for $i = 1, \ldots, n$. Carlitz (1953) gave the first constructive proof of Theorem 1.21 using constructions from an earlier proof by Vijayaraghavan and Chowla (1948), that the multiplicative group of units in the ring of integers modulo n ($n > 2$) admits a complete mapping if and only if it is not cyclic. The proof given here is derived from Carlitz' proof.

For non-abelian groups the Hall-Paige conjecture remains unsolved. Hall and Paige (1955) did prove their conjecture true for several classes of groups; solvable groups; S_n, the symmetric group on n symbols; and A_n, the alternating group on n symbols. Orthomorphisms of many other groups can be be constructed from orthomorphisms of these groups using the quotient group construction of Corollary 1.15.

In the remainder of this section we will prove Hall and Paige's existence results and list more recent results, due to Aschbacher, Dénes and Keedwell, Evans, Saeli, and Di Vincenzo. First we will note a fairly easy existence proof.

Theorem 1.22. Any group of odd order admits orthomorphisms.

Proof. Consider the mapping $\theta: x \rightarrow x^2$. To solve the equation $\theta(x) = y$ for x, let m be the order of y and set $x = y^{(m+1)/2}$. Thus θ is a permutation. The mapping $x \rightarrow x^{-1}\theta(x) = x$ is clearly also a permutation and hence θ is an orthomorphism. ∎

In proving that any finite group of odd order can be the multiplicative group of a neofield Paige (1949) showed $x \rightarrow x$ to be a complete mapping of any finite group of odd order. Sade (1963) again proved the mapping $x \rightarrow x$ to be a complete mapping of any finite group of odd order as well as any commutative quasigroup of odd order.

Theorem 1.23. Hall and Paige (1955, Theorem 1). Let H be a subgroup of a finite group G and let U be a two-sided set of coset representatives for H in G. If H admits an orthomorphism and there exists a permutation σ of U such that $\{u\sigma(u): u \in U\}$ is a left-sided set of coset representatives of H in G then G admits an orthomorphism.

Proof. For $u \in U$ and $h \in H$ we know that $\sigma(u)h = h*u'$ for some $h* \in H$ and $u' \in U$, where $h*$ and u' are uniquely determined by u and h. Let θ be an orthomorphism of H and define $\theta': G \rightarrow G$ by $\theta'(uh*) = u\sigma(u)\theta(h)$. Then θ' is an orthomorphism of G. $\blacksquare$

Corollary 1.14. Hall and Paige (1955, Corollary 1). Let G be a factorable group, i.e G has nontrivial subgroups A and B such that $G = AB$ and $A \cap B = \{e\}$. If A and B both admit orthomorphisms then so does G.

Proof. In Theorem 1.23 set $A = U, B = H$, and $\sigma(u) = u^{-1}\theta(u)$, θ and orthomorphism of A. $\blacksquare$

Corollary 1.15. Hall and Paige (1955). If $H \triangleleft G$ and H and G/H admit orthomorphisms then G admits orthomorphisms also.

Proof. Let U be a two-sided set of coset representatives for H in G and let ϕ be a complete mapping of G/H. Define $\sigma: U \rightarrow U$ by $\phi(uH) = \sigma(u)H$. Then σ, U, and H satisfy the conditions of Theorem 1.23. $\blacksquare$

Lemma 1.3. Let $H = <a>$, a an involution in $Z(G)$ and let G/H admit orthomorphisms. If there exists $b \in G$, b of order a power of 2, $b^i \neq a$ for any even integer i, then G admits orthomorphisms.

Proof. In this proof we will construct complete mappings of G from complete mappings of G/H. As groups admit orthomorphisms if and only if they admit complete mappings, this will prove the lemma. Let $\phi: G \rightarrow G/H$ be the canonical homomorphism and let θ be a complete mapping of G/H.

Define a graph Γ, whose vertices are the elements of G, x being adjacent to y if $y = xa, xb$, or xb^{-1}. Γ is bipartite and regular. Choose $\mathcal{K}$ and $\mathcal{K}'$ to be bipartite classes of Γ (This choice need not be unique as Γ need not be connected).

Define $\phi': G/H \rightarrow \mathcal{K}$ by $\phi'(y) = x$ if $\phi(x) = y$ and $x \in \mathcal{K}$. Let $K = \{x: x\phi'\theta\phi(x) \in \mathcal{K}\}$ and $K' = \{x: x\phi'\theta\phi(x) \in \mathcal{K}'\}$. Then K and K' partition G.

Next form a new graph Γ', whose vertices are the elements of K, x being adjacent to y if $\phi'\theta\phi(x)ab = \phi'\theta\phi(y)$ or vice versa. Now Γ' is a regular bipartite graph and so we may choose bipartite classes A and B of Γ'.

Finally define $\theta': G \rightarrow G$ by $\theta'(x) = \phi'\theta\phi(x)$ if $x \in A$, $\phi'\theta\phi(x)b$ if $x \in B$, $\phi'\theta\phi(x)ab$ if $x \in Aa$, and $\phi'\theta\phi(x)a$ if $x \in Ba$. This is a complete mapping of G. $\blacksquare$

Lemma 1.4. Hall and Paige (1955, Theorem 4). Any non-abelian, non-cyclic 2-group admits orthomorphisms.

Proof. Assume G to be a minimal counterexample and let u be an involution in $Z(G)$. Then $G/<u>$ is non-cyclic, as otherwise G would be abelian. Now G cannot contain another involution as Lemma 1.3 would then imply that G admitted orthomorphisms. Thus

G must be a generalized quaternion group. To complete the proof we will show that all generalized quaternion groups admit orthomorphisms. Specifically, for $n \geq 3$, let $m = 2^{n-2}$ and $G = <a, b| \ a^{2m} = e, b^2 = a^m, bab^{-1} = a^{-1}> = \{a^i b^j: i = 0, \dots, 2m - 1, j = 0, 1\}$. Define $\theta: G \to G$ as follows.

$$\theta(a^i) = a^{2i} \quad \text{, for } i = 0, \dots, m - 1$$
$$\theta(a^i) = a^{2i-m}b \quad \text{, for } i = m, \dots, 2m - 1$$
$$\theta(a^i b) = a^{2i+1}b \quad \text{, for } i = 0, \dots, m - 1$$
$$\theta(a^i b) = a^{2i+1} \quad \text{, for } i = m, \dots, 2m - 1$$

Then θ is an orthomorphism of G. ∎

Hall and Paige indicated that this result was already known. Paige (1951) credits Bruck with having proved this and claims that arguing modulo the commutator subgroup enables one to establish this result for those 2-groups with non-cyclic commutator subgroups and that Bruck had dealt with the remaining cases. Hall and Paige first constructed complete mappings for dihedral 2-groups, generalized quaternion groups, and semi-dihedral 2-groups, and then argued inductively, using maximal subgroups, to establish the existence of complete mappings for all non-cyclic 2-groups. The proof given here uses Hall and Paige's construction of complete mappings of generalized quaternion groups and a previously unpublished lemma of Evans (Lemma 1.3).

Theorem 1.24. Hall and Paige (1955, Theorem 6). Any solvable group whose Sylow 2-subgroup is non-cyclic admits orthomorphisms.
Proof. By a theorem of Philip Hall any solvable group is factorable. In fact if G is solvable then $G = AB$ where $|A|$ is odd and B is a Sylow 2-subgroup of G. The result then follows from Lemma 1.4 and Corollary 1.14. ∎

Theorems 1.21 and 1.22 are in fact special cases of this result, as is the following corollary.

Corollary 1.16. C. P. Johnson (1981). Dihedral groups of doubly even order admit complete mappings.

Theorem 1.25. Hall and Paige (1955, Theorem 2). S_n, the symmetric group on n symbols, admits orthomorphisms for $n > 3$.
Proof. The proof is by induction on n. S_4 is solvable with a non-cyclic Sylow 2-subgroup and so admits orthomorphisms by Theorem 1.24. It is thus sufficient to show how orthomorphisms of S_n, $n > 4$ can be constructed from orthomorphisms of S_{n-1} to complete the proof. The set $U = \{\text{identity}, (1, n), (2, n), \dots, (n - 1, n)\}$ is a two-sided set of coset representatives of S_{n-1} in S_n and the permutation σ of U defined by $\sigma(\text{identity}) = \text{identity}$, $\sigma((i, n)) = (i + 1, n)$ for $i = 2, \dots, n - 2$ and $\sigma((n - 1, n)) = (2, n)$ satisfies the hypothesis of Theorem 1.23. The result then follows. ∎

Theorem 1.26. Hall and Paige (1955, Theorem 3). A_n, the alternating group on n symbols, admits orthomorphisms for all values of n.

Proof. For $n \leq 4$, A_n is a solvable group with a trivial or non-cyclic Sylow 2-subgroup and so admits orthomorphisms by Theorem 1.24. To establish the result for $n > 4$ it suffices to show how orthomorphisms of A_n can be constructed from orthomorphisms of A_{n-1}. The set $U = \{$identity, $(1, n-1, n)$, $(1, n, n-1)$, $(2, n)(1, n-1)$, $(3, n)(1, n-1)$, ... , $(n-2, n)$ $(1, n-1)\}$ is a two-sided set of coset representatives of A_{n-1} in A_n and the permutation σ of U defined by $\sigma(\text{identity}) = \text{identity}$, $\sigma((1, n-1, n)) = (1, n-1, n)$, $\sigma((1, n, n-1)) = (1, n, n-1)$, $\sigma((i, n)(1, n-1)) = (i+1, n)(1, n-1)$ for $i = 2, ... , n-3$ and $\sigma((n-2, n)(1, n-1)) = (2, n)(1, n-1)$ satisfies the hypothesis of Theorem 1.23. Hence the result. ∎

Very little progress was made towards resolving the Hall-Paige conjecture until quite recently. Since 1989 the existence of complete mappings has been established for several new classes of groups. In the following theorems we list these new results without proofs.

Theorem 1.27. Saeli (1989). $SL(2, q)$ admits complete mappings if $q = 2^n > 2$.

Theorem 1.28. Di Vincenzo (1989). $Sz(2^{2n+1})$ admits complete mappings if $n \geq 1$.

Theorem 1.29. Di Vincenzo (1989). $SU(3, q^2)$ and $PSU(3, q^2)$ admit complete mappings if q is even.

Theorem 1.30. Evans (to appear - c). $GL(2, q)$ and $SL(2, q)$ admit complete mappings if $q = 2^n > 2$.

Hall and Paige (1955) proved that, with a technical condition, if the Sylow 2-subgroups of a finite group G are non-cyclic and intersect pairwise trivially then G admits complete mappings. In the course of removing this condition Di Vincenzo (1989) proved Theorems 1.28 and 1.29. The result of Theorem 1.30 has since been extended, in unpublished work of Evans, to show that $GL(n, q)$, $PGL(n, q)$, $SL(n, q)$, and $PSL(n, q)$ admit complete mappings, for $n \geq 2$ and q even, $q \neq 2$. These proofs and the proofs of Theorems 1.28, 1.29, and 1.30 depend on appropriate partitions of groups.

Dénes and Keedwell (1989) point out that any finite non-solvable group must have a non-cyclic Sylow 2-subgroup. They conjecture that all finite non-solvable groups admit complete mappings, a conjecture that is equivalent to the Hall-Paige conjecture. They also prove that, when suitably ordered, the product of all the elements of a finite non-solvable group is the identity. This is a necessary condition for a group to admit complete mappings.

The most significant recent advances toward solving the Hall-Paige conjecture are due to Aschbacher. If this conjecture were false, which groups could be minimal counterexamples. Aschbacher showed these to be "essentially" simple groups, and went on to restrict the classes of simple groups that could occur in these minimal counterexamples. His results were presented in a series of lectures given at NSA in the summer of 1990.

Theorem 1.31. (Aschbacher's reduction) Let G be a minimal counterexample to Hall and Paige's conjecture. Then G has a normal subgroup L such that $L/Z(L)$ is simple and $Z(L)$, $C_G(L)$, and G/L are cyclic 2-groups.

The subgroup L of Theorem 1.27 is a quasisimple group, and if L is simple then G is an almost simple group. Let H be a subgroup of a finite group G and let U be a two-sided set of coset representatives for H in G. If there exists a bijections $\sigma, \rho: U \to U$ such that $u\sigma(u)H = \rho(U)H$ then Aschbacher calls $\{H, U; \sigma, \rho\}$ an *HP-system* for G. By Theorem 1.23 if G admits an HP-system $\{H, U; \sigma, \rho\}$ and H admits an orthomorphism then so does G. Aschbacher conjectures the following, which he proves implies the Hall-Paige conjecture.

HP-system conjecture. Any finite almost simple group G admits an HP-system (H, D, ζ, η) with H a proper subgroup of G, the Sylow 2-subgroup of H being non-trivial and non-cyclic.

Theorem 1.32. The HP-system conjecture implies the Hall-Paige conjecture.

Theorem 1.33. The HP-system conjecture holds for A_n and S_n.

Theorem 1.33 is actually a consequence of Theorems 1.25 and 1.26.
The last three theorems of this sections, also due to Aschbacher, establish the truth of the HP-system conjecture for many classes of almost simple groups.

Theorem 1.34. Let G be an almost simple group with minimal normal subgroup L of Lie type over $GF(q)$, G/L a cyclic 2-group. If the HP-system conjecture does not hold for G then one of the following must be true.
i) L is of Lie rank 1, q is odd, and G induces inner diagonal automorphisms on L.
ii) $L = L_n(q)$, $n \geq 3$, G nontrivial on the Dynkin diagram of L.
iii) $L = Sp_4(q)$ or $F_4(q)$, q even, G nontrivial on the Dynkin diagram of L.
iv) $L = {}^2F_4(q)$, q even.

Theorem 1.35. If G is an almost simple group, with minimal normal subgroup a Mathieu group, then G satisfies the HP-system conjecture.

Chapter 2: Elementary abelian groups

Section 1. Introduction.

Any elementary abelian group can be thought of as the additive group of a finite field, enabling us to use multiplication in the field to construct orthomorphisms. In particular the mappings $x \to ax$, $a \neq 0, 1$, form a $(q - 2)$ - clique of $\text{Orth}(GF(q)^+)$. Hence $\omega(\text{Orth}(GF(q)^+)) = q - 2$. This is the only class of groups which is known to admit a complete set of orthomorphisms. In this and the next chapter we will be concerned with the structure of orthomorphism graphs of elementary abelian groups.

Orthomorphism graphs of $GF(q)^+$ were first studied in the search for non-Desarguesian affine planes of prime order. Thus the case q a prime was the first case to be studied. However the methods used to construct and study orthomorphisms and orthomorphism graphs of Z_p generalized fairly easily to elementary abelian groups.

It has long been conjectured that all affine (or equivalently projective) planes of prime order must be Desarguesian. This conjecture has been proved for orders 2, 3, 5, and 7 only. A weaker form of this conjecture would be that any affine plane of prime order, admitting translations, must be Desarguesian. As such a plane must be a Cartesian plane, by Corollary 1.7 any affine plane of prime order p, admitting translations, must correspond to a $(p - 2)$ - clique of $\text{Orth}(Z_p)$. The following theorem, due to Evans and McFarland (1984), gives a rationale for the study of the structure $\text{Orth}(Z_p)$.

Theorem 2.1. A non-Desarguesian affine plane of prime order p, admitting translations, exists if and only if $\text{Orth}(Z_p)$ admits more than one $(p - 2)$ - clique.
Proof. As any affine plane of prime order, admitting translations, must be a Cartesian plane by Corollary 1.7, any affine plane of prime order p, admitting translations, corresponds to a $(p - 2)$ - clique of $\text{Orth}(Z_p)$. Also, by Corollary 1.5, any $(p - 2)$ - clique of $\text{Orth}(Z_p)$ can be used to construct an affine plane of order p. Thus, to prove the theorem, it is sufficient to prove that the Desarguesian plane can only be constructed from the clique $\{x \to ax: a \neq 0,1\}$. Let $\theta_2, \ldots, \theta_{p-1}$ be a clique of $\text{Orth}(Z_p)$. We may assume without loss of generality that $\theta_i(1) = i$. Let us further suppose that the corresponding affine plane is Desarguesian. Then the result of Theorem 1.16 v) implies that $\theta_i \in \text{Aut}(G)$, which implies that $\theta_i(x) = ix$ for all x. Hence the result. ∎

A similar result does not hold for prime powers as $\text{Orth}(GF(8)^+)$ contains eight 6-cliques, but there exists only one affine plane of order 8. This will be shown in Section 2 of Chapter 6.

Corollary 2.1. A non-Desarguesian affine plane of order 11 or 13 cannot admit translations.
Proof. For $p = 11$ this was first proved by Johnson, Dulmage, and Mendelsohn (1961). It has since been reproved by Cates and Killgrove (1981) and by Evans and McFarland (1984).

For $p = 13$ this was first proved by Cates and Killgrove (1981), and later reproved by Mendelsohn and Wolk (1985). All of these proofs resulted from computer searches. ∎

The result of Corollary 2.1 has not been extended and the problem of the existence of non-Desarguesian affine planes of prime order, admitting translations, remains unsolved. Computer generated results exist for the structure of elementary abelian groups of small order and our study is also motivated in part by the problem of explaining this data. In this regard we will be partially successful.

Some elementary counting results follow.

Theorem 2.2. Let p be a prime, $p > 3$.

i) The number of vertices of $\text{Orth}(Z_p)$ is congruent to p - 2 modulo p.

ii) The number of vertices of $\text{Orth}(Z_p)$ of degree 0 is congruent to 0 modulo p.

iii) The number of vertices of $\text{Orth}(Z_p)$ of degree r, not of the form $x \rightarrow ax$, is congruent to 0 modulo p.

iv) The number of edges of $\text{Orth}(Z_p)$ is congruent to 3 modulo p.

v) The number of $(p$ - 2$)$ - cliques of $\text{Orth}(Z_p)$ is congruent to 1 modulo p.

vi) The degree of $x \rightarrow ax$ is congruent to p - 3 modulo p.

Proof. This follows immediately from the fact that, if $g \neq 0$ then T_g has order p and fixes an orthomorphism if and only if the orthomorphism is of the form $x \rightarrow ax$. ∎

The results of this theorem hold even if the only orthomorphisms of Z_p are of the form $x \rightarrow ax$. Thus this theorem by itself does not give us much information about the structure of $\text{Orth}(Z_p)$.

In constructing and studying orthomorphisms and orthomorphism graphs of elementary abelian groups we will adopt three different approaches. In this f we will construct and study orthomorphisms by solving systems of linear equations over finite fields, and in chapter 3 using quadratic residues, or more generally cyclotomy, and permutation polynomials.

Section 2. A linear algebra approach.

Let us consider again the class of orthomorphisms of the form $x \rightarrow ax$. We wish to modify this construction to yield other classes of orthomorphisms. One modification due to Evans (1987b) is as follows. Let θ be an orthomorphism of $\text{GF}(q)^+$ and let $x_1, \ldots, x_{q-1}$ be the non-zero elements of $\text{GF}(q)$. Then for any $a \neq 0, 1$ we can write $\theta(x_i) = ax_{\sigma(i)}$ for a uniquely determined permutation σ of $\{1, \ldots, q - 1\}$ and $\theta(x_i) - x_i = (a - 1)x_{\varepsilon(i)}$ for another uniquely determined permutation ε of $\{1, \ldots, q - 1\}$. Should $\sigma(i) = i$ then

$$\theta(x_i) - x_i = ax_{\sigma(i)} - x_i = (a - 1)x_i = (a - 1)x_{\varepsilon(i)} \text{ and so } \varepsilon(i) = i. \text{ Similarly if } \varepsilon(i) = i$$

then $\sigma(i) = i$ and if $\varepsilon(i) = \sigma(i)$ then $\varepsilon(i) = \sigma(i) = i$.

Thus, in principle, all orthomorphisms of $GF(q)^+$ can be found by solving systems of linear equations of the form $ax_{\sigma(i)} - x_i = (a - 1)x_{\varepsilon(i)}$, where $a \neq 0, 1$, and σ and ε are permutations of $\{1, \dots, q - 1\}$ for which $\varepsilon(i) = i$ if and only if $\sigma(i) = i$ if and only if $\varepsilon(i) = \sigma(i)$.

In the above we have constructed systems of linear equations from orthomorphisms. We will now give an algorithm for constructing classes of orthomorphisms from systems of linear equations.

Step 1. Pick σ and ε, permutations of $\{1, \dots, q - 1\}$ for which $\varepsilon(i) = i$ if and only if $\sigma(i) = i$ if and only if $\varepsilon(i) = \sigma(i)$.

Step 2. Solve the system of equations $ax_{\sigma(i)} - x_i = (a - 1)x_{\varepsilon(i)}$, $i = 1, \dots, q - 1$, for $a, x_1, \dots, x_{q-1}$.

Step 3. If we can find a solution in which $a \neq 0, 1$ and $x_1, \dots, x_{q-1}$ are all the non-zero elements of $GF(q)$ then define θ by $\theta(x_i) = ax_{\sigma(i)}$ and $\theta(0) = 0$. This will be an orthomorphism of $GF(q)^+$.

We shall refer to the system of equations $ax_{\sigma(i)} - x_i = (a - 1)x_{\varepsilon(i)}$, $i = 1, \dots, q - 1$, $\sigma(i) = i$ if and only if $\varepsilon(i) = i$ if and only if $\sigma(i) = \varepsilon(i)$, as a (σ, ε) - *system* and we shall say that a (σ, ε) - system *yields orthomorphisms* if there is a solution $a, x_1, \dots, x_{q-1}$ of the system for which the mapping $x_i \rightarrow ax_{\sigma(i)}$ and $0 \rightarrow 0$ is an orthomorphism. We shall call $|\{i \colon \sigma(i) \neq i\}|$ the *order* of the system. (σ, ε) - systems of orders 1 and 2 do not exist. Of course different (σ, ε) - systems may yield the same set of orthomorphisms. For instance, this will occur if one can be obtained from the other by a relabelling of $x_1, \dots, x_{q-1}$. We shall say that a (σ, ε) - system is *equivalent* to a (σ', ε') - system if there exists a permutation ϕ of the set $\{1, \dots, q - 1\}$ for which $\sigma' = \phi^{-1}\sigma\phi$ and $\varepsilon' = \phi^{-1}\varepsilon\phi$. Equivalent systems yield the same set of orthomorphisms. For simplicity we will not write down equations of the form $ax_i - x_i = (a - 1)x_i$ as, once we have a solution of the restricted system $ax_{\sigma(i)} - x_i = (a - 1)x_{\varepsilon(i)}$, $i \in \{j \colon \varepsilon(j) \neq j\}$, for which the x_i are all distinct and non-zero, then the corresponding orthomorphism θ is defined by setting $\theta(x_i) = ax_{\sigma(i)}$ if $\varepsilon(i) \neq i$ and ax_i otherwise. By the term, *rank of the solution space* of a (σ, ε) - system, we shall actually mean the rank of the solution space of the restricted system $ax_{\sigma(i)} - x_i = (a - 1)x_{\varepsilon(i)}$, $i \in \{j \colon \varepsilon(j) \neq j\}$.

In the above discussion we could replace $GF(q)^+$ by R^+, where R is a commutative ring with unity. The modifications that this entails are fairly minor. For this to work we do require that there exist $a \in R$ for which a and $a - 1$ are both units of R. In all the examples presented in this chapter, we will deal with both the special case $GF(q)^+$ as well as the general case R^+.

Let us consider a simple example.

Example 2.1. Suppose $\sigma = (1\ 2\ 3)$ and $\varepsilon = (1\ 3\ 2)$. Then the corresponding system of equations is:

$$ax_2 - x_1 = (a - 1)x_3$$
$$ax_3 - x_2 = (a - 1)x_1$$
$$ax_1 - x_3 = (a - 1)x_2$$

This is in fact, up to equivalence, the only system of order 3. We have the constant solution $x_1 = x_2 = x_3$. Hence the solution space has rank at least 1 and this system will yield orthomorphisms only if the solution space has rank at least 2. Thus the determinant of any 2×2 principal submatrix of the coefficient matrix cannot be a unit.

If we are performing our calculations in a field then this means that $a^2 - a + 1 = 0$. In this case our system will only yield orthomorphisms if $a \neq 0$, 1 and $a^2 - a + 1 = 0$. Solving for x_3 in terms of x_1 and x_2 and setting $c = x_1$ and $b = (x_2 - x_1)/(a - 1)$ we find that $(x_1, x_2, x_3) = b(0, a - 1, a) + c(1, 1, 1)$. Thus x_1, x_2, and x_3 will be distinct and non-zero if and only if $c, c + b(a - 1), c + ba$ are all distinct and non-zero.

If these conditions are satisfied then the corresponding orthomorphism acts as follows.

c	$\rightarrow$	$a(b(a - 1) + c)$
$b(a - 1) + c$	$\rightarrow$	$a(ab + c)$
$ab + c$	$\rightarrow$	ac
x	$\rightarrow$	ax, for all other values of x.

This construction gives us orthomorphisms of $GF(q)^+$ if and only if 3 divides $q - 1$. This system is a special case of a class of (σ, ε)-systems that will be studied in more detail in the next section.

If we are performing our calculations in a commutative ring with unity then, for the calculations to remain valid, we require a and $a - 1$ to be units. The conditions we obtained for this system to yield orthomorphisms must then be modified. Now $a^2 - a + 1$ cannot be a unit. In fact, we find that $(a^2 - a + 1)(x_2 - x_1) = 0$ and so $(a^2 - a + 1)b = 0$. The condition that x_1, x_2, x_3 be distinct and non-zero becomes the condition that c, $b(a - 1) + c$, and $ab + c$ be distinct and non-zero. If all these conditions are satisfied then this system yields orthomorphisms and the description of the orthomorphisms is the same as in the field case.

As an example, we obtain orthomorphisms of Z_{15} by setting $a = 2$, 8, or 14, $b = 5$ or 10, and choosing any value of c other than 0, 5, or 10.

Section 3. A special class of examples.

In Example 2.1, $\sigma = (1\ 2\ 3)$ and $\varepsilon = \sigma^2$. We can generalize this to $\sigma = (1\ 2\ \dots\ n)$ and

$\varepsilon = \sigma^r$. In this section we will study systems of this type, restricting ourselves to fields. Our results are taken from Evans (1987b). The first result establishes the existence of solutions to these systems whenever n divides q - 1.

Theorem 2.3. Let n be a divisor of q - 1 and let $2 \leq r < n \leq q$ - 1. Let $b, c \in$ GF(q), $b \neq 0$, c not an nth. root of unity. Let g be a primitive nth. root of unity and let $a = (1 - g^r)/(g - g^r)$. Then the mapping $\theta(x) = ab(g^{i+1} - c)$ if $x = b(g^i - c)$, ax otherwise, is an orthomorphism of GF(q)$^+$.

Proof. Define permutations ε and σ of $\{1, \ldots, q - 1\}$ by $\sigma = (\,1\,2\,\ldots\,n)$ and $\varepsilon = \sigma^r$. Set $x_i = b(g^i - c)$ for $i \leq n$. Then $ax_{\sigma(i)} - x_i = (a - 1)x_{\varepsilon(i)}$ is a (σ, ε) - system and $a, x_1, \ldots, x_n$ is a solution to this system. Also $a \neq 0, 1$ and $x_1, \ldots, x_n$ are distinct and non-zero. ∎

Given a (σ, ε) - system in which $\sigma = (1\,2\,3\,\ldots\,n)$ and $\varepsilon = \sigma^r$, must n necessarily divide q - 1 for the system to yield orthomorphisms. Further, when n does divide q - 1 are the only solutions possible those given in Theorem 2.3. We can give only partial answers to these questions.

Theorem 2.4. Let n and r be positive integers, $2 \leq r < n$. Let $\sigma = (1\,2\,3\,\ldots\,n)$ and $\varepsilon = \sigma^r$. If the solution space of this system has rank at most two for any choice of a then the system yields orthomorphisms if and only if n divides q - 1. In this case the orthomorphisms obtained are precisely those obtained in Theorem 2.3.

Proof. Let P denote the $n \times n$ permutation matrix $\begin{bmatrix} 0 & 1 & & \\ & & \ddots & \\ & & & 1 \\ 1 & & & 0 \end{bmatrix}$ and let

$A = aP - I + (1 - a)P^r$. If $X = (x_1, \ldots, x_n)^T$ then the system can be written as $AX = 0$. Let F be the nth. cyclotomic extension of GF(q) and let g be a primitive nth. root of unity in F. Then the eigenvalues of A are $\{\lambda_i = ag^i - 1 + (1 - a)g^{ir}: i = 0, \ldots, n - 1\}$ and the eigenvector of A corresponding to λ_i is $b(g^i, g^{2i}, \ldots, 1)^T$, $b \neq 0$.

Now $n - 2 \leq \text{rank } A \leq n - 1$ and to obtain a solution for $a, x_1, \ldots, x_n$ with $a \neq 0, 1$ and $x_i \neq$ constant we must have rank $A = n - 2$. This can only occur if for some i, $a(g^i - g^{ir}) = 1 - g^{ir}$ and $x_j = b(g^i)^j + c$. The mapping $\theta: x_i \rightarrow ax_{\sigma(i)}$ can be an orthomorphism only if $x_1, \ldots, x_n$ are distinct, i. e. $b \neq 0$ and g^i a primitive nth. root of unity. Also θ can be an orthomorphism of GF(q)$^+$ only if $x_j \in$ GF(q), $j = 1, \ldots, n$, from which it follows that $g^i \in$ GF(q). Thus n divides q - 1 and the result then follows. ∎

If $r = 2$ or n - 1 then the solution space has rank at most two and hence the result of Theorem 2.4 holds. The question of whether the solution space is always of rank two or less remains open.

Section 4. Systems of order 5 or less.

In this section we list and solve all inequivalent (σ, ε) - systems of order 5 or less, with the exception of the special class of systems dealt with in the previous section. We will solve these systems over fields and over commutative rings with unity. We will leave out many of the calculations as these are similar to the calculations of Example 2.1.

Example 2.2. Suppose $\sigma = \varepsilon = $ identity. The order of this system is 0 and the corresponding orthomorphisms are precisely the mappings $x \to ax$, where in a field we need the condition that $a \neq 0, 1$, and in a ring the condition that a and $a - 1$ be units.

Example 2.3. Suppose $\sigma = (1\ 2)(3\ 4)$ and $\varepsilon = (1\ 3)(2\ 4)$. Then the corresponding system of equations is:

$$ax_2 - x_1 = (a - 1)x_3$$
$$ax_1 - x_2 = (a - 1)x_4$$

$$ax_4 - x_3 = (a - 1)x_1$$
$$ax_3 - x_4 = (a - 1)x_2$$

We find that $2(x_2 - x_1) = 0$ and so 2 cannot be a unit. In a field this implies that q is a power of 2, $q \geq 8$. Solving for x_3 and x_4 in terms of x_1 and x_2 and setting $b = (x_2 - x_1)/(a - 1)$ and $c = x_1$ we can write $(x_1, x_2, x_3, x_4) = b(0, a - 1, a, -1) + c(1, 1, 1, 1)$. In a field this defines an orthomorphism if and only if $a \neq 0, 1$ and $c, c + b(a - 1), c + ba$, and $c - b$ are all distinct and non-zero. In a ring this defines an orthomorphism if and only if a and $a - 1$ are units, $2b = 0$, and $c, c + b(a - 1), c + ba$, and $c - b$ are all distinct and non-zero.

If these conditions are satisfied then the corresponding orthomorphism acts as follows.

c	$\to$	$a(ab - b + c)$
$ab - b + c$	$\to$	ac
$ab + c$	$\to$	$a(c - b)$
$c - b$	$\to$	$a(ab + c)$
x	$\to$	ax, for all other values of x.

Example 2.4. Suppose $\sigma = (1\ 2)(3\ 4)$ and $\varepsilon = (1\ 3\ 2\ 4)$. Then the corresponding system of equations is:

$$ax_2 - x_1 = (a - 1)x_3$$
$$ax_1 - x_2 = (a - 1)x_4$$

$$ax_4 - x_3 = (a - 1)x_2$$
$$ax_3 - x_4 = (a - 1)x_1$$

We find that $(a^2 + 1)(x_1 - x_2) = 0$ and so $a^2 + 1$ cannot be a unit. In a field this means that $a^2 + 1 = 0$ which implies that q is odd, $q > 7$, $q \equiv 1$ modulo 4.

Setting $c = x_1$ and $b = (x_2 - x_1)/(a - 1)$ we can write the solution as $(x_1, \ldots, x_4) = b(0, a - 1, a, -1) + c(1, 1, 1, 1)$. In a field this defines an orthomorphism if and only if $a \neq 0, 1, a^2 + 1 = 0$, and $c, c + b(a - 1), c + ba$, and $c - b$ are all distinct and non-zero. In a ring this defines an orthomorphism if and only if a and $a - 1$ are units, $(a^2 + 1)b = 0$, and $c, c + b(a - 1), c + ba$, and $c - b$ are all distinct and non-zero.

If these conditions are satisfied then the corresponding orthomorphism acts as follows.

c	$\rightarrow$	$a(ab - b + c)$
$ab - b + c$	$\rightarrow$	ac
$ab + c$	$\rightarrow$	$a(c - b)$
$c - b$	$\rightarrow$	$a(ab + c)$
x	$\rightarrow$	ax, for all other values of x.

Note that the description of the orthomorphism is identical to the description of the orthomorphism in Example 2.3. These results differ only in the conditions imposed on a, b, and c.

Example 2.5. Suppose $\sigma = (1\ 2)(3\ 4\ 5)$ and $\varepsilon = (1\ 3\ 5\ 2\ 4)$. Then the corresponding system of equations is:

$$ax_2 - x_1 = (a - 1)x_3$$
$$ax_1 - x_2 = (a - 1)x_4$$

$$ax_4 - x_3 = (a - 1)x_5$$
$$ax_5 - x_4 = (a - 1)x_1$$
$$ax_3 - x_5 = (a - 1)x_2$$

We find that $(2a^2 - a + 1)(x_2 - x_1) = 0$ and so $2a^2 - a + 1$ cannot be a unit. In a field

q must be odd, $q \geq 7$, $2x^2 - x + 1$ must factor in $GF(q)$ and $2a^2 - a + 1 = 0$.

Then setting $c = x_1$ and $b = (x_2 - x_1)/(a - 1)^2$ we can write the solution as

$(x_1, \ldots, x_5) = b(0, (a-1)^2, a(a-1), 1-a, -2a) + c(1, \ldots, 1)$. In a field this defines an orthomorphism if and only if $a \neq 0, 1$, $2a^2 - a + 1 = 0$, and c, $c + b(a-1)^2$, $c + ba(a-1)$, $c + b(1-a)$, and $c - 2ba$ are all distinct and non-zero. In a ring this defines an orthomorphism if and only if a and $a - 1$ are units, $(2a^2 - a + 1)b = 0$, and c, $c + b(a-1)^2$, $c + ba(a-1)$, $c + b(1-a)$, and $c - 2ba$ are all distinct and non-zero.

If these conditions are satisfied then the corresponding orthomorphism acts as follows.

$$
\begin{array}{ccl}
c & \rightarrow & a(b(a-1)^2 + c) \\
b(a-1)^2 + c & \rightarrow & ac \\
ab(a-1) + c & \rightarrow & a(b(1-a) + c) \\
b(1-a) + c & \rightarrow & a(c - 2ab) \\
c - 2ab & \rightarrow & a(ab(a-1) + c) \\
x & \rightarrow & ax, \text{ for all other values of } x.
\end{array}
$$

Example 2.6. Suppose $\sigma = (1\ 2)(3\ 4\ 5)$ and $\varepsilon = (1\ 3\ 5)(2\ 4)$. Then the corresponding system of equations is:

$$ax_2 - x_1 = (a-1)x_3$$
$$ax_1 - x_2 = (a-1)x_4$$

$$ax_4 - x_3 = (a-1)x_5$$
$$ax_5 - x_4 = (a-1)x_2$$
$$ax_3 - x_5 = (a-1)x_1$$

We find that $(a^2 - a + 2)(x_2 - x_1) = 0$ and so $a^2 - a + 2$ cannot be a unit. In a field q must be odd, $q \geq 7$, $x^2 - x + 2$ must factor in $GF(q)$ and $a^2 - a + 2 = 0$.

Then setting $c = x_1$ and $b = (x_2 - x_1)/(a-1)^2$ we can write the solution as

$(x_1, \ldots, x_5) = b(0, (a-1)^2, a(a-1), 1-a, -2a) + c(1, \ldots, 1)$. In a field this defines an orthomorphism if and only if $a \neq 0, 1$, $a^2 - a + 2 = 0$, and c, $c + b(a-1)^2$, $c + ba(a-1)$, $c + b(1-a)$, $c - 2ba$ are all distinct and non-zero. In a ring this defines an orthomorphism if and only if a and $a - 1$ are units, $(a^2 - a + 2)b = 0$, and c, $c + b(a-1)^2$, $c + ba(a-1)$, $c + b(1-a)$, $c - 2ba$ are all distinct and non-zero.

If these conditions are satisfied then the corresponding orthomorphism acts as follows.

c	$\rightarrow$	$a(b(a-1)^2+c)$
$b(a-1)^2+c$	$\rightarrow$	ac
$ab(a-1)+c$	$\rightarrow$	$a(b(1-a)+c)$
$b(1-a)+c$	$\rightarrow$	$a(c-2ab)$
$c-2ab$	$\rightarrow$	$a(ab(a-1)+c)$
x	$\rightarrow$	ax, for all other values of x.

Note that the description of the orthomorphism is identical to the description of the orthomorphism in Example 2.5. These results differ only in the conditions imposed on a, b, and c.

Example 2.7. Suppose $\sigma = (1\ 2\ 3\ 4\ 5)$ and $\varepsilon = (1\ 3\ 5\ 4\ 2)$. Then the corresponding system of equations is:

$$ax_2 - x_1 = (a-1)x_3$$
$$ax_3 - x_2 = (a-1)x_1$$
$$ax_4 - x_3 = (a-1)x_5$$
$$ax_5 - x_4 = (a-1)x_2$$
$$ax_1 - x_5 = (a-1)x_4$$

This system yields no orthomorphisms in a field as from the first two equations we determine that $(a^2 - a + 1)(x_2 - x_1) = 0$ and so $a^2 - a + 1 = 0$. From the last two equations we determine that $a^2 x_1 - (a-1)x_2 = (a^2 - a + 1)x_4 = 0$. Thus the solution space of this system of equations has rank 1 and so the only solutions are constant solutions. However in a ring we can easily establish that $(x_1, \ldots, x_5) = $
$b(0, a(a^2 - a + 1), a^2 - a + 1, 1, 1 - a) + c(1, 1, 1, 1, 1)$, where $a^2 - a + 1$ is not a unit, $b(a^2 - a + 1) \neq 0$ but $b(a^2 - a + 1)^2 = 0$.

The corresponding orthomorphism acts as follows.

c	$\rightarrow$	$a(ba(a^2 - a + 1) + c)$
$ba(a^2 - a + 1) + c$	$\rightarrow$	$a(b(a^2 - a + 1) + c)$
$b(a^2 - a + 1) + c$	$\rightarrow$	$a(b+c)$
$b+c$	$\rightarrow$	$a(b(1-a)+c)$
$b(1-a)+c$	$\rightarrow$	ac
x	$\rightarrow$	ax, for all other values of x.

As an example, consider the ring of integers modulo 45 and pick $a = 2$, $b = 5$, and $c = 1$.

Example 2.8. Suppose $\sigma = (1\ 2\ 3\ 4\ 5)$ and $\varepsilon = (1\ 3)(2\ 5\ 4)$. Then the corresponding system of equations is:

$$ax_2 - x_1 = (a - 1)x_3$$
$$ax_3 - x_2 = (a - 1)x_5$$
$$ax_4 - x_3 = (a - 1)x_1$$
$$ax_5 - x_4 = (a - 1)x_2$$
$$ax_1 - x_5 = (a - 1)x_4$$

We find that $(2a^2 - 3a + 2)(x_2 - x_1) = 0$ and so $2a^2 - 3a + 2$ cannot be a unit. In a field q must be odd, $q \geq 7$, $2x^2 - 3x + 2$ must factor in $GF(q)$ and $2a^2 - 3a + 2 = 0$.

Then setting $c = x_1$ and $b = (x_2 - x_1)/(a - 1)^2$ we can write the solutions as

$(x_1, \ldots, x_5) = b(0, (a - 1)^2, a(a - 1), a - 1, a^2 - a + 1) + c(1, \ldots, 1)$. In a field this defines an orthomorphism if and only if $a \neq 0, 1$, $2a^2 - 3a + 2 = 0$, and $c, c + b(a - 1)^2$, $c + ba(a - 1)$, $c + b(a - 1)$, and $c + b(a^2 - a + 1)$ are all distinct and non-zero. In a ring this defines an orthomorphism if and only if a and $a - 1$ are units, $(2a^2 - 3a + 2)b = 0$, and $c, c + b(a - 1)^2, c + ba(a - 1), c + b(a - 1)$, and $c + b(a^2 - a + 1)$ are all distinct and non-zero.

If these conditions are satisfied then the corresponding orthomorphism acts as follows.

c	$\rightarrow$	$a(b(a - 1)^2 + c)$
$b(a - 1)^2 + c$	$\rightarrow$	$a(ab(a - 1) + c)$
$ab(a - 1) + c$	$\rightarrow$	$a(b(a - 1) + c)$
$b(a - 1) + c$	$\rightarrow$	$a(b(a^2 - a + 1) + c)$
$b(a^2 - a + 1) + c$	$\rightarrow$	ac
x	$\rightarrow$	ax, for all other values of x.

Chapter 3: Cyclotomic orthomorphisms

Section 1. Linear and quadratic orthomorphisms.

Any orthomorphism of GF(q)$^+$ can be written in the form $x_i \to ax_{\sigma(i)}$, where $\{x_1, \ldots, x_{q-1}\} = GF(q)^*$ and $a \neq 0$. In Chapter 2 we used this description to show how orthomorphisms could be constructed by solving systems of linear equations. While we were able to construct classes of orthomorphisms in this manner, this approach yields little information about the structure of orthomorphism graphs, even for orthomorphism graphs induced by the solutions to a given (σ, ε) - system.

We will consider a variation of this approach, based on the fact that any orthomorphism of GF(q)$^+$ can also be written in the form $x_i \to a_i x_i$, where $\{x_1, \ldots, x_{q-1}\} = GF(q)^*$ and $a_i \neq 0$ for all i. Let $X_1, \ldots, X_r$ be a partition of the set of elements of GF(q)* and define $\theta(x) = b_i x$ if $x \in X_i$. For what values of $b_1, \ldots, b_r$ is θ an orthomorphism and when are two such orthomorphisms adjacent? We will show that for certain partitions these questions have fairly simple answers. As an example, suppose we choose X_1 to be the set of non-zero squares of GF(q) and X_2 to be the set of non-squares of GF(q). Then, for $q = 7$, setting $b_1 = 3$ and $b_2 = 5$ defines an orthomorphism of Z_7, and setting $b_1 = 5$ and $b_2 = 3$ defines another orthomorphism of Z_7. In fact, any orthomorphism of Z_7 that is not of the form $x \to ax$ is a translate of one of these two orthomorphisms.

Let us first consider the simplest partition of all, the partition consisting of only one set. This partition defines mappings of the form $x \to ax$, which we shall denote by the symbol $[a]$. We know that $[a]$ is a permutation if and only if $a \neq 0$, and an orthomorphism if and only if $a \neq 0, 1$. We shall use $\mathfrak{C}_1$ or $\mathfrak{C}_1(q)$ to denote the set of orthomorphisms $\{[a]: a \neq 0, 1\}$ as well as the corresponding induced orthomorphism graph, and we shall call the elements of $\mathfrak{C}_1$ *linear orthomorphisms*. The elements of $\mathfrak{C}_1$ form a complete set of orthomorphisms of GF(q)$^+$.

The action of congruences on $\mathfrak{C}_1$ is easily determined. If $\theta \in \mathfrak{C}_1$ then $T_g[\theta] = \theta$ for all g. For q prime this characterizes linear orthomorphisms. $H_\alpha[\theta] = \theta$ for all $\alpha \in \mathfrak{C}_1$ if and only if $\theta \in \mathfrak{C}_1$. For the congruences I and R, $R[[a]] = [1 - a]$ and $I[[a]] = [1/a]$. The action of I and R on $\mathfrak{C}_1$ is summarized in the following diagram.

$$[\,\vartheta\,] \quad \overset{R}{\leftrightarrow} \quad [\,1-\vartheta\,] \quad \overset{I}{\leftrightarrow} \quad [\,1/(1-\vartheta\,)\,]$$

$$\updownarrow I \qquad\qquad\qquad\qquad\qquad \downarrow R$$

$$[\,1/\vartheta\,] \quad \leftrightarrow \quad [\,1-(1/\vartheta\,)\,] \quad \leftrightarrow \quad [\,\vartheta/(\,\vartheta-1\,)\,]$$
$$\qquad\quad R \qquad\qquad\qquad I$$

In view of the fact that for $\theta \in \mathfrak{C}_1$, $H_\alpha[\theta] = \theta$ for all $\alpha \in \mathfrak{C}_1$, and $T_g[\theta] = \theta$ for all g it might be asked: What other classes of orthomorphisms can be found that are fixed by congruences? All the classes of orthomorphisms constructed in this, and the next, chapter can be defined in terms of the congruences that fix them.

For q odd, the non-zero elements of GF(q) can be partitioned into two sets, the non-zero squares and the non-squares. This defines mappings of the form $x \to Ax$ if x is a non-zero square, or Bx if x is a non-square, which we shall denote by the symbol $[A, B]$. We shall use $\mathfrak{C}_2$ or $\mathfrak{C}_2(q)$ to denote the set of such mappings that are orthomorphisms as well as the induced orthomorphism graph, and we shall call the elements of $\mathfrak{C}_2$ *quadratic orthomorphisms*. Note that $\mathfrak{C}_1(q) \subsetneq \mathfrak{C}_2(q)$. Quadratic orthomorphisms were first defined and studied by Mendelsohn and Wolk (1985) and subsequently by Evans (1987a).

Theorem 3.1. Let $\theta = [A, B]$ and $\phi = [C, D]$.

i) θ is a permutation if and only if A/B is a non-zero square.

ii) $\theta \sim \phi$ if and only if $(A - C)/(B - D)$ is a non-zero square.

iii) θ is an orthomorphism if and only if A/B and $(A - 1)/(B - 1)$ are both non-zero squares.

Proof.

i)   If θ is a permutation and A is a non-zero square then θ maps non-zero squares to non-zero squares and hence non-squares to non-squares and so B must also be a non-zero square. The case A a non-square and the converse are proved similarly.

ii)   This follows from i) by noting that, if $\delta(x) = \theta(x) - \phi(x)$ then $\delta = [A - C, B - D]$.

iii)   This follows from i) and ii) by setting $C = D = 1$. ∎

Two immediate corollaries.

Corollary 3.1.

i) $[A, B]$ is an orthomorphism if and only if $[B, A]$ is an orthomorphism.

ii) $[A, B] \sim [C, D]$ if and only if $[B, A] \sim [D, C]$.

iii)   If $A \neq B$ then $[A, B] \sim [B, A]$ if and only if -1 is a square.

Corollary 3.2. Let $\alpha = [a] \in \text{Aut}(\text{GF}(q)^+)$. Then $H_\alpha[A, B] = [A, B]$ if a is a square and $[B, A]$ if a is a non-square.

Corollary 3.2 embodies a characterization of quadratic orthomorphisms.

Theorem 3.2. Let $\theta \in \text{Orth}(GF(q)^+)$. Then $H_\alpha[\theta] = \theta$ for all $\alpha \in \{[a^2]: a \neq 0\}$ if and only if $\theta \in \mathfrak{S}_2$.

Proof. The if part of this proof was established in Corollary 3.2. For the only if part suppose that $H_\alpha[\theta] = \theta$ for all $\alpha \in \{[a^2]: a \neq 0\}$ and that c is a non-zero square, $\theta(c) = bc$, and $\alpha(x) = a^2x$. Then $H_\alpha[\theta](a^2c) = ba^2c$ and so $\theta(x) = bx$ for all squares x. The case c a non-square is similar. ∎

The actions of the congruences I, R, and T_g on $\mathfrak{S}_2$ are listed in the next theorem.

Theorem 3.3.

i) $I[[A, B]] = [1/A, 1/B]$ if A is a non-zero square and $[1/B, 1/A]$ if A is a non-square.

ii) $R[[A, B]] = [1 - A, 1 - B]$ if -1 is a square and $[1 - B, 1 - A]$ if -1 is a non-square.

iii) If $A \neq B$ and $g \neq 0$ then $T_g[[A, B]] \notin \mathfrak{S}_2$.

Proof. The proofs of i) and ii) are routine. For iii) suppose, for $g \neq 0$, that $T_g[[A, B]] = [C, D]$. As $\mathfrak{S}_2(q) = \mathfrak{S}_1(q)$ for $q \leq 5$ we may assume that $q \geq 7$. If g is a square then there exists a non-zero square x for which $x + g$ is a non-zero square and a non-square y for which $y + g$ is a non-zero square. Then $Cx = A(x + g) - Ag = Ax$ and $Dy = A(y + g) - Ag = Ay$. But then $C = D = A$ and so $[A, B](x) = T_{-g}T_g[[A, B]](x) = T_{-g}[[A]](x) = [A](x)$. A contradiction. The case g a non-square is similar. ∎

It is fairly easy to generate all quadratic orthomorphisms for q a prime. Write down the elements of $GF(q)^*$ in order and mark the squares. The values of A and B that correspond to orthomorphisms can then be read off by inspection, as A and B must either both be marked or both be unmarked, and $A - 1$ and $B - 1$ must also either both be marked or both be unmarked.

Example 3.1. We list the elements of $GF(7)^*$ with the squares marked.

$$1 \; \overset{\cdot}{2} \; 3 \; \overset{\cdot}{4} \; 5 \; 6$$

We obtain 5 orthomorphisms from $A = B = 2, \ldots , 6$. For $A \neq B$ we note that the only allowed choices are $A = 3, B = 5$, or $A = 5, B = 3$.

Example 3.2. We list the elements of $GF(11)^*$ with the squares marked.

$$1 \; 2 \; \overset{\cdot}{3} \; \overset{\cdot}{4} \; \overset{\cdot}{5} \; 6 \; 7 \; 8 \; \overset{\cdot}{9} \; 10$$

We obtain 9 orthomorphisms from $A = B = 2, \ldots , 10$. For $A \neq B$ we find 12 orthomorphisms corresponding to the following values of A and B.

A	2	2	3	4	6	7	D
B	6	10	9	5	10	8	A

Thus $GF(11)^+$ has 21 quadratic orthomorphisms, 9 of which are linear.

Example 3.3. We list the elements of $GF(13)^*$ with the squares marked.

$$1 \ 2 \ 3 \ 4 \ 5 \ 6 \ 7 \ 8 \ 9 \ 10 \ 11 \ 12$$

We obtain 11 orthomorphisms from $A = B = 2, \dots, 12$. For $A \neq B$ we find 20 orthomorphisms corresponding to the following values of A and B.

A	2	2	3	3	4	5	6	6	7	9	B
B	5	11	9	12	10	11	7	8	8	12	A

Thus $GF(13)^+$ has 31 quadratic orthomorphisms, 11 of which are linear.

In general we can give a formula for the number of quadratic orthomorphisms for any value of q. This formula is derived from the following theorem that gives us a method for generating all non-linear, quadratic orthomorphisms.

Theorem 3.4. Evans (1987a). Let R and S be non-zero squares, $R \neq S$, $R, S \neq 1$, and define $A = S(1 - R)/(S - R)$ and $B = (1 - R)/(S - R)$. Then $[A, B]$ is an orthomorphism and all non-linear, quadratic orthomorphisms are of this form.
Proof. $A/B = S$ and $(A - 1)/(B - 1) = R$ if and only if $A = S(1 - R)/(S - R)$ and $B = (1 - R)/(S - R)$. If $A/B = S$ and $(A - 1)/(B - 1) = R$ then $R = S$ if and only if $A = B$, $R = 1$ if and only if $A = B$, and $S = 1$ if and only if $A = B$. ∎

Corollary 3.3. $|\mathfrak{C}_2(q)| = (q - 3)(q - 5)/4 + (q - 2)$.

A different proof of this, using cyclotomy numbers, is given in Mendelsohn and Wolk (1985).

Theorem 3.5. Evans (1987a). Suppose $A \neq B$, and let L be a non-zero square, $L \neq 1$, A/B, $(A - 1)/(B - 1)$. Define $C = (LB - A)/(L - 1)$. Then $[A, B] \sim [C]$ and all linear orthomorphisms adjacent to $[A, B]$ can be generated in this way.
Proof. $(A - C)/(B - C) = L$ if and only if $C = (LB - A)/(L - 1)$. If $(A - C)/(B - C) = L$ then $L = A/B$ if and only if $C = 0$, and $L = (A - 1)/(B - 1)$ if and only if $C = 1$. ∎

Corollary 3.4. If $A \neq B$ then $[A, B]$ is adjacent to precisely $(q - 7)/2$ linear orthomorphisms.

Again for the special case q a prime there is an easy way to determine which linear orthomorphisms a given quadratic orthomorphism is adjacent to. As before write down the elements of $GF(q)^*$ in order and mark the squares. Then $[A, B] \sim [C]$ if and only if the numbers C spaces before A and C spaces before B are either both marked or both unmarked.

Example 3.4. As in Example 3.2, we list the elements of $GF(11)^*$ with the squares marked.

$$1 \; 2 \; 3 \; 4 \; \dot{5} \; 6 \; 7 \; 8 \; \dot{9} \; 10$$

Let us consider the quadratic orthomorphism $[2, 6]$. By inspection we see that $[2, 6] \sim [C]$, a linear orthomorphism, if and only if $C = 7$ or 8.

Theorem 3.6. Evans (1987a). Let $[A, B]$ be a non-linear, quadratic orthomorphism and suppose $[C, D] \neq [A, B]$. If either $C/D = A/B$ or $(C - 1)/(D - 1) = (A - 1)/(B - 1)$ then $[A, B] \sim [C, D]$.
Proof. By routine calculations we can show that if $C/D = A/B$ then $(A - C)/(B - D) = C/D = A/B$ and if $(C - 1)/(D - 1) = (A - 1)/(B - 1)$ then $(A - C)/(B - D) = (C - 1)/(D - 1) = (A - 1)/(B - 1)$. ∎

From this we obtain a lower bound on the degree of a non-linear, quadratic orthomorphism.

Corollary 3.5. The degree of a non-linear, quadratic orthomorphism is at least $(3q - 21)/2$.
Proof. Let $[A, B]$ be a non-linear quadratic orthomorphism. By Corollary 3.4, $[A, B]$ is adjacent to $(q - 7)/2$ linear orthomorphisms. Next we will count the number of quadratic orthomorphisms $[C, D]$, $C/D = A/B$ and $A \neq C$. If $[C, D]$ is a quadratic orthomorphism, $C/D = A/B$ and $A \neq C$, then $[C, D]$ is non-linear and $(C - 1)/(D - 1) \neq (A - 1)/(B - 1)$. By Theorem 3.6, each such orthomorphism is adjacent to $[A, B]$. Let R be a square, $R \neq 1$, A/B, $(A - 1)/(B - 1)$. Then, by Theorem 3.4, each non-linear, quadratic orthomorphism $[C, D]$, $C/D = A/B$ and $A \neq C$, can be found by solving $C/D = A/B$ and $(C - 1)/(D - 1) = R$ for C and D. This yields a further $(q - 7)/2$ neighbors of $[A, B]$. Similarly we find that the number of non-linear, quadratic orthomorphisms $[C, D]$, for which $(C - 1)/(D - 1) = (A - 1)/(B - 1)$, $A \neq C$, is also equal to $(q - 7)/2$. These two latter sets are disjoint as $C/D = A/B$ and $(C - 1)/(D - 1) = (A - 1)/(B - 1)$ if and only if $C = A$ and $D = B$. Hence the degree of $[A, B]$ is at least $3(q - 7)/2$. ∎

For $q = 11$ this shows that each non-linear, quadratic orthomorphism has degree at least 6. In fact, for $q = 11$, each non-linear, quadratic orthomorphism has degree 8. If $q = 7$ then each non-linear, quadratic orthomorphism has degree 0. It is tempting to expect that in general each non-linear, quadratic orthomorphism has the same number of neighbors in $\mathfrak{G}_2$. While this holds true for $q = 9$ it fails to hold for $q = 13$. In this case, 12 elements of $\mathfrak{C}_2 - \mathfrak{C}_1$ have 12 neighbors in $\mathfrak{C}_2$, 6 have 14 neighbors in $\mathfrak{C}_2$, and 2 have 16 neighbors in $\mathfrak{C}_2$.

Another question of interest. How many $(q - 2)$-cliques can we find in $\mathfrak{C}_2(q)$? We have clearly shown the answer to be 1 for $q = 3, 5$, and 7. In fact the answer is 1 if q is a prime. This was established for $q = 13$ and 17 by Mendelsohn and Wolk (1985) using a computer and for $q \leq 47$ by Evans (1987a) using simple hand calculations. The complete answer to this question relies on the theory of permutation polynomials and will be given in Chapter 5, Section 5. If q is not a prime then the answer is at least 2. To see this suppose

$q = p^r, p$ a prime, $r \neq 1$, and let $t \mid q$. Then the orthomorphisms $[A, A^t]$, $A \neq 0, 1$, form a $(q - 2)$-clique of $\mathfrak{C}_2(q)$. In fact, each $t = 1, p, \ldots, p^{r-1}$ yields a distinct $(q - 2)$-clique. This has immediate consequences for the case $q = 9$. By Corollary 3.3, $\mathfrak{C}_2(9) - \mathfrak{C}_1(9)$ has 6 elements. By Corollary 3.4 each of these orthomorphisms is adjacent to exactly one element of $\mathfrak{C}_1(9)$ and so each orthomorphism of $\mathfrak{C}_2(9) - \mathfrak{C}_1(9)$ must be adjacent to the same orthomorphism of $\mathfrak{C}_1(9)$ as well as to each other. More concrete data for $q = 9$ is presented in the next example.

Example 3.5. GF(9) can be thought of as consisting of terms of the form $a + bi, a, b \in$ GF(3), i a solution of $x^2 + 1 = 0$, and addition and subtraction modulo 3. The squares are 1, 2, i, and $2i$, and the non-squares are $1 + i, 1 + 2i, 2 + i$, and $2 + 2i$.

Let us determine the quadratic orthomorphisms $[A, B]$. We obtain 7 orthomorphisms from $A = B$. For $A \neq B$ we find 6 quadratic orthomorphisms corresponding to the following values of A and B.

$$A \mid \quad i \qquad 1 + i \qquad 2 + i \quad \mid B$$

$$B \mid \quad 2i \qquad 1 + 2i \qquad 2 + 2i \quad \mid A$$

Each of these orthomorphisms is adjacent to exactly one element of $\mathfrak{C}_1$. In fact they are all adjacent to [2] and the elements of $\mathfrak{C}_2 - \mathfrak{C}_1$ together with [2] form a 7-clique of Orth(GF(9)$^+$). This clique corresponds to the dual translation plane of order 9. See Theorem 1.16 ii).

Section 2. Cyclotomic orthomorphisms.

For q odd, the non-zero squares form a subgroup of GF(q)* of index 2. To generalize the partition used in Section 1, let H be a subgroup of GF(q)* of index e and use the cosets $H = C_0, \ldots, C_{e-1}$ of H as the classes of the partition. These can be described quite explicitly. Let g be a primitive element of GF(q), $q = ef + 1$. Then we define C_i to be the set $\{g^{ej+i}: j = 0, \ldots, f - 1\}, i = 0, \ldots, e - 1$. We call C_i the ith *cyclotomy class* (with respect to e) of GF(q). A further definition will be needed. The number of solutions to $x_i + 1 = x_j$, $x_i \in C_i$, $x_j \in C_j$, is denoted (i, j). These numbers are called *cyclotomy numbers*. We define a class of mappings θ: GF(q) $\rightarrow$ GF(q) by $\theta(0) = 0$ and $\theta(x) = A_i x$ if $x \in C_i$, and we use $\mathfrak{C}_e$ or $\mathfrak{C}_e(q)$ to denote both the set of such mappings that are orthomorphisms as well as the corresponding induced orthomorphism graph. We shall use $[A_0, \dots, A_{e-1}]$ to denote the mapping θ: GF(q) $\rightarrow$ GF(q) defined by $\theta(x) = A_i x$ if $x \in C_i$, $\theta(0) = 0$, and we shall call the elements of $\mathfrak{C}_e$ *cyclotomic orthomorphisms* (*of index e*).

Theorem 3.7.
i) $[A_0, \dots, A_{e-1}]$ is a permutation if and only if the mapping $C_i \rightarrow A_i C_i$ permutes the cyclotomy classes.

ii) $[A_0, \dots, A_{e-1}] \sim [B_0, \dots, B_{e-1}]$ if and only if the mapping $C_i \rightarrow (A_i - B_i)C_i$ permutes the cyclotomy classes.

iii) $[A_0, \dots, A_{e-1}]$ is an orthomorphism if and only if the mappings $C_i \rightarrow A_i C_i$ and $C_i \rightarrow (A_i - 1)C_i$ both permute the cyclotomy classes.
Proof. Similar to the proof of Theorem 3.1. ∎

Example 3.6. If $e = 1$ then there is only one cyclotomy class and this consists of all the non-zero elements of GF(q). Thus $\mathfrak{C}_1(q) = \{[a]: a \neq 0, 1\}$, the class of linear orthomorphisms.

Example 3.7. At the other extreme if $e = q - 1$ then there are precisely $q - 1$ cyclotomy classes each consisting of one element only. Thus $\mathfrak{C}_{q-1}(q) = \text{Orth}(\text{GF}(q)^+)$.

Example 3.8. If q is odd then $2 \mid (q - 1)$ and the cyclotomy classes corresponding to $e = 2$ consist of the non-zero squares and the non-squares. This defines the quadratic orthomorphisms.

Theorem 3.8. Let $\theta = [A_0, \dots, A_{e-1}] \in \mathfrak{C}_e$ and let $\alpha(x) = ax$, $a \in C_k$.
i) $H_\alpha[\theta] \in \mathfrak{C}_e$ and $H_\alpha[\theta](x) = A_{i-k}x$ for $x \in C_i$.

ii) $R[\theta] \in \mathfrak{C}_e$ and $R[\theta](x) = (1 - A_{i+m})x$ for $x \in C_i$ and $-1 \in C_m$.

iii) $I[\theta] \in \mathfrak{C}_e$ and $I[\theta](x) = (1/A_i)x$ for $x \in A_i C_i$.
Proof. This follows from routine calculations. ∎

Can we obtain a result analogous to Theorem 3.3 iii)? That is, is it true that if $\theta \in \mathfrak{C}_e - \mathfrak{C}_1$, $e \neq q - 1$, then $T_g[\theta] \notin \mathfrak{C}_e$ if $g \neq 0$? The answer is yes, but the proof requires the use of permutation polynomials and will be given in Section 4 of this chapter.

Theorem 3.9. $\theta \in \mathfrak{C}_e$ if and only if $H_\alpha[\theta] = \theta$ for all $\alpha \in \{[a^e]: a \neq 0\}$.
Proof. As $a^e \in C_0$ the only if part has already been proved in Theorem 3.8 i). To prove

the if part assume that $H_\alpha[\theta] = \theta$ for all $\alpha \in \{[a^e]: a \neq 0\}$. If $x, y \in C_i$ then $y = a^e x$
for some a and so $\alpha(x) = y = a^e x$. If $\theta(x) = bx$ then $H_\alpha[\theta](y) = H_\alpha[\theta](a^e x) =$
$a^e ba^{-e} a^e x = by$. The result follows. ∎

Theorem 3.10. Let $a, b \mid (q - 1)$. If $a|b$ then $\mathfrak{C}_a \subseteq \mathfrak{C}_b$.
Proof. If $\theta \in \mathfrak{C}_a$ then $H_\alpha[\theta] = \theta$ for all $\alpha \in \{[c^a]: c \neq 0\}$ by Theorem 3.9. But
$\{[c^b]: c \neq 0\} \subseteq \{[c^a]: c \neq 0\}$ and so $H_\beta[\theta] = \theta$ for all $\beta \in \{[c^b]: c \neq 0\}$. Hence
$\theta \in \mathfrak{C}_b$ by Theorem 3.9. ∎

In fact the sets $\mathfrak{C}_e$ form a lattice which is isomorphic to a sublattice of the subgroup
lattice of Z_{q-1}. These two lattices need not be isomorphic in general as some of these sets of
orthomorphisms may be identical. As an example, if $q = 5$ then $\mathfrak{C}_4 = \mathfrak{C}_2 = \mathfrak{C}_1$. Another
example: If $q = 7$ then $\mathfrak{C}_6 \neq \mathfrak{C}_2 \neq \mathfrak{C}_3 = \mathfrak{C}_1$.

Theorem 3.11. (Evans (1989c)). Suppose $q = ef + 1$. Then

$$|\mathfrak{C}_e| = \sum \prod_{i=0}^{e-1} (\alpha(i) - i, \beta(i) - i),$$

where subtraction is modulo e and the summation runs over all ordered pairs of permutations
α, β of the set $\{0, \ldots, e - 1\}$.
Proof. $[A_0, \ldots, A_{e-1}]$ is an orthomorphism if and only if $A_i C_i = C_{\beta(i)}$ and $(A_i - 1)C_i$
$= C_{\alpha(i)}$ for some pair of permutations α, β of the set $\{0, \ldots, e - 1\}$. Thus
$A_i \in C_{\beta(i) - i} \cap (C_{\alpha(i) - i} + 1)$, where the indices are subtracted modulo e. Thus the
number of orthomorphisms $[A_0, \ldots, A_{e-1}]$ satisfying $A_i C_i = C_{\beta(i)}$ and $(A_i - 1)C_i =$
$C_{\alpha(i)}$ is equal to $\prod_{i=0}^{e-1} (\alpha(i) - i, \beta(i) - i)$. Thus the sum over all pairs of permutations α, β
of the set $\{0, \ldots, e - 1\}$ counts the number of elements of $\mathfrak{C}_e$. ∎

Example 3.9. If $e = 1$ then there is only one cyclotomy class, $(0, 0) = q - 2$, and
$|\mathfrak{C}_1(q)| = q - 2$.

Example 3.10. If $e = 2$ then there are two cyclotomy classes and four ordered pairs of
permutations to consider. Using information in Storer (1967), we know that if $(q - 1)/2$ is
odd then $(0, 0) = (1, 0) = (1, 1) = (q - 3)/4$ and $(0, 1) = (q + 1)/4$, and if $(q - 1)/2$ is even
then $(0, 1) = (1, 0) = (1, 1) = (q - 1)/4$ and $(0, 0) = (q - 5)/4$. Thus $|\mathfrak{C}_2(q)| =$
$(0, 0)^2 + (0, 1)^2 + (1, 0)^2 + (1, 1)^2 = (q^2 - 4q + 7)/4 = (q - 3)(q - 5)/4 + |\mathfrak{C}_1(q)|$, and we
have obtained another proof of Corollary 3.3. This is the proof given in Mendelsohn and
Wolk (1985).

For $e = 3$, the formula of Theorem 3.11 can be written as follows.

$$|\mathfrak{C}_3| = 3\left[\sum_{j=0}^{2}\prod_{i=0}^{2}(i,j) + \sum_{i=0}^{2}\prod_{j=0}^{2}(i,j) + \sum_{k=0}^{2}\prod_{i=0}^{2}(i,i+k)\right] + \sum_{i,j=0}^{2}(i,j)^3$$

Thus we can easily calculate $|\mathfrak{C}_3|$, knowing the entries of the cyclotomic matrix, the 3×3 matrix with ijth entry the cyclotomy number (i, j).

Example 3.11. For $q = 13$, the cyclotomy classes are $C_0 = \{1, 8, 12, 5\}$, $C_1 = \{2, 3, 11, 10\}$, and $C_2 = \{4, 6, 9, 7\}$, and the cyclotomic matrix is $\begin{bmatrix} 0 & 1 & 2 \\ 1 & 2 & 1 \\ 2 & 1 & 1 \end{bmatrix}$. From this we find that $|\mathfrak{C}_3(13)| = 65$.

Example 3.12. For $q = 16$, let α be a root of the primitive polynomial $x^4 + x^3 + 1 = 0$. Then the cyclotomy classes are $C_0 = \{1, \alpha^3, \alpha^3 + \alpha^2 + \alpha + 1, \alpha^2 + 1, \alpha + 1\}$, $C_1 = \{\alpha, \alpha^3 + 1, \alpha^2 + \alpha + 1, \alpha^3 + \alpha, \alpha^2 + \alpha\}$, and $C_2 = \{\alpha^2, \alpha^3 + \alpha + 1, \alpha^3 + \alpha^2 + \alpha, \alpha^3 + \alpha^2 + 1, \alpha^3 + \alpha^2\}$, and the cyclotomic matrix is $\begin{bmatrix} 0 & 2 & 2 \\ 2 & 2 & 1 \\ 2 & 1 & 2 \end{bmatrix}$. From this we find that $|\mathfrak{C}_3(16)| = 122$.

Example 3.13. For $q = 19$, the cyclotomy classes are $C_0 = \{1, 8, 7, 18, 11, 12\}$, $C_1 = \{2, 16, 14, 17, 3, 5\}$, and $C_2 = \{4, 13, 9, 15, 6, 10\}$, and the cyclotomic matrix is $\begin{bmatrix} 2 & 1 & 2 \\ 1 & 2 & 3 \\ 2 & 3 & 1 \end{bmatrix}$. From this we find that $|\mathfrak{C}_3(19)| = 233$.

Example 3.14. Terms in the formula for $|\mathfrak{C}_3|$ can actually be used to construct elements of $\mathfrak{C}_3$. As an example, suppose $q = 13$ and consider the term $(k, m)^3$ in the formula for $|\mathfrak{C}_3|$. If $[A_0, A_1, A_2]$ is an orthomorphism corresponding to this term then $A_i \in C_k$ and $A_i + 1 \in C_m$. Thus if we pick $k = m = 1$ then $A_i \in \{2, 10\}$ and each of the 8 choices for A_0, A_1, and A_2 yields an orthomorphism.

Section 3: Permutation polynomials.

It is known that every function from $GF(q)$ to $GF(q)$ can be represented by a polynomial of degree at most $q - 1$. To see this, first note that the number of such functions is q^q, which equals the number of polynomials over $GF(q)$ of degree at most $q - 1$. Further, no two distinct polynomials can represent the same function as then their difference would be a nonzero polynomial of degree at most $q - 1$ that has each element of $GF(q)$ as a root. We shall call those polynomials that represent permutations, *permutation polynomials*; those that represent orthomorphisms of the additive group, *orthomorphism polynomials*; and those that represent complete mappings of the additive group, *complete mapping polynomials*. A polynomial $f(x)$, for which $f(0) = 0$, will be an orthomorphism polynomial if and only if both $f(x)$ and $f(x) - x$ are permutation polynomials, and a complete mapping polynomial if and only if both $f(x)$ and $f(x) + x$ are permutation polynomials. For more information on permutation polynomials see Lidl and Niederreiter (1983, Chapter 7). One useful characterization of permutation polynomials follows.

Theorem 3. 12. (Hermite's criterion). Let $f(x) = a_n x^n + \ldots + a_1 x + a_0$. Then $f(x)$ is a permutation polynomial of GF(q) if and only if the coefficient of x^{q-1} in the reduction of $f(x)^t$ modulo $x^q - x$ is zero for $t = 1, \ldots, q - 2$, t relatively prime to q, and nonzero for $t = q - 1$.
Proof. See Lidl and Niederreiter (1983, Chapter 7). ∎

As an immediate corollary.

Corollary 3.6. The degree of a permutation polynomial, reduced modulo $x^q - x$, is at most $q - 2$.

In fact for orthomorphisms and complete mappings we can do better. The following improvement was proved by Niederreiter and Robinson (1982) for q odd, and by Wan (1986) for q even.

Theorem 3.13. The degree of an orthomorphism or complete mapping polynomial, reduced modulo $x^q - x$, is at most $q - 3$ for $q > 2$.
Proof. If q is odd and $\theta(x)$ is an orthomorphism polynomial of GF(q) of degree $q - 2$ then $(\theta(x) - x)^2$ will have the same degree modulo $x^q - x$ as $2\theta(x)x$, which has degree $q - 1$, an impossibility.

If q is even then GF(q) $\cong E/2E$ for some ring E of algebraic integers. Let $\eta: E \rightarrow$ GF(q) be the canonical homomorphism. Then η can be extended to a homomorphism from the polynomial ring $E[x]$ to the polynomial ring GF(q)[x] by setting $\eta(x) = x$. Let g be a primitive element of GF(q) and let $h \in E$ satisfy $\eta(h) = g$. Then certainly $h^{q-1} \equiv 1$ modulo 2 and $h^i \not\equiv 1$ modulo 2 for $1 \le i < q - 1$. We may assume without loss of generality that $h^{q-1} \equiv 1$ modulo 4 as should this not be the case then we can replace h by h^q. This works because $\eta(h^q) = g$ and $(h^q)^{q-1} =$
$(h(1 + 2((h^{q-1} - 1)/2)))^{q-1} \equiv h^{q-1} + 2h^{q-1}(q - 1)((h^{q-1} - 1)/2) \equiv$
$1 + 2((h^{q-1} - 1)/2) - 2h^{q-1}((h^{q-1} - 1)/2) \equiv 1$ modulo 4.

Set $S = \{h^i: 1 \leq i \leq q - 1\} \cup \{0\}$ and let $f(x)$ be any permutation polynomial of GF(q) and $F(x)$ a polynomial over E for which $\eta(F(x)) = f(x)$. Then $\{\eta(s): s \in S\} = \{\eta(F(s)): s \in S\} = $ GF(q) and so $\sum_{s \in S} F(s)^2 = \sum_{s \in S} (s + 2G(s))^2$ for some function $G: E \to E$. Hence $\sum_{s \in S} F(s)^2 \equiv \sum_{s \in S} s^2$ (modulo 4) $= h^2((h^{q-1})^2 - 1)/(h^2 - 1) \equiv 0$ modulo 4.

Thus if $f(x)$ is an orthomorphism polynomial of GF(q) and $\eta(F(x)) = f(x)$ then $\sum_{s \in S} F(s)^2 \equiv 0$ modulo 4 and $\sum_{s \in S} (F(s) - s)^2 \equiv 0$ modulo 4. Hence $2\sum_{s \in S} sF(s) \equiv 0$ modulo 4. Suppose that $f(x) = a_{q-2}x^{q-2} + a_{q-3}x^{q-3} + \dots + a_0$ and $F(x) = b_{q-2}x^{q-2} + b_{q-3}x^{q-3} + \dots + b_0$. Then as $\sum_{s \in S} s^i \equiv 0$ modulo 2 for $i = 1, \dots, q - 2$, and $\sum_{s \in S} s^{q-1} \equiv 0$, $2b_{q-2} \equiv 0$ modulo 4 and so $a_{q-2} = \eta(b_{q-2}) = 0$ proving that the degree of $f(x)$ is at most $q - 3$.

The case $\theta(x)$ a complete mapping polynomial follows immediately as $\theta(x)$ is a complete mapping polynomial if and only if $\theta(x) + x$ is an orthomorphism polynomial. ∎

If the polynomial $f(x)$ is a permutation polynomial then so is the polynomial $h(x) = af(x + b) + c$, where $a \neq 0$. We can choose a, b and c so that $h(x)$ is monic, has constant term 0, and should its degree n be relatively prime to q then the coefficient of x^{n-1} is zero. We shall call such a permutation polynomial a *normalized permutation polynomial*. Clearly to determine all permutation polynomials of degree n it is sufficient to determine all normalized permutation polynomials of degree n. Note that if θ is an orthomorphism polynomial then there exists a normalized permutation polynomial f, and $a, b \in$ GF(q) such that $\theta(x) = aT_b[f](x)$ and f is adjacent to the permutation $x \to a^{-1}x$. Also if f is a normalized permutation polynomial adjacent to the permutation $x \to a^{-1}x$ then for all $b \in$ GF(q) the polynomial $aT_b[f](x)$ will be an orthomorphism polynomial of GF(q).

A complete list of the normalized permutation polynomials of degree at most 5 can be found in Dickson (1897), Lidl and Niederreiter (1983) and in Niederreiter and Robinson (1982) and a list of normalized permutation polynomials of degree 6 can be found in Dickson (1897). All normalized permutation polynomials of degree 5 or less are given in Theorem 3.14 a). A partial list of normalized permutation polynomials of degree 6 for q odd is given in Theorem 3.14 b). All orthomorphism polynomials of degree 5 or less are given in Theorem 3.14 a), and all orthomorphism polynomials of degree 6 for q relatively prime to 6 are given in Theorem 3.15 b).

Theorem 3.14.

a)   The only normalized permutation polynomial of degree less than 6 are:

i) x, for all q

ii) x^2, if and only if $q \equiv 0$ modulo 2

iii) x^3, if $q \not\equiv 1$ modulo 3

iv) $x^3 - ax$, if a is not a square and $q \equiv 0$ modulo 3

v) $x^4 \pm 3x$, if $q = 7$

vi) $x^4 + ax^2 + bx$, if $x = 0$ is the only root in $GF(q)$ and $q \equiv 0$ modulo 2

vii) x^5, if $q \not\equiv 1$ modulo 5

viii) $x^5 - ax$, if a is not a fourth power and $q \equiv 0$ modulo 5

ix) $x^5 + ax$, if $a^2 = 2$ and $q = 9$

x) $x^5 \pm 2x^2$, if $q = 7$

xi) $x^5 + ax^3 \pm x^2 + 3a^2x$, if a is not a square and $q = 7$

xii) $x^5 + ax^3 + 5^{-1}a^2x$, if $q \equiv \pm 2$ modulo 5

xiii) $x^5 + ax^3 + 3a^2x$, if a is not a square and $q = 13$

xiv) $x^5 - 2ax^3 + a^2x$, if a is not a square and $q \equiv 0$ modulo 5.

b) For q odd the following are normalized permutation polynomials:

i) $x^6 \pm 2x$, $q = 11$

ii) $x^6 \pm 4x$, $q = 11$

iii) $x^6 + a^2x^3 + ax^2 + 5x$, a a square, $q = 11$

iv) $x^6 - a^2x^3 + ax^2 - 5x$, a a square, $q = 11$

v) $x^6 + 4a^2x^3 + ax^2 + 4x$, a a nonsquare, $q = 11$

vi) $x^6 - 4a^2x^3 + ax^2 - 4x$, a a nonsquare, $q = 11$

vii) $x^6 + a^2x^4 + bx^3 + a^4x^2 + (2a^2b + a^5)x$, a square, $a \neq 0$, $b = 0$, $\sqrt{2}a^3$, a^3, $(\sqrt{2} + 1)a^3$, $q = 9$

viii) $x^6 + a^2x^4 + bx^3 + a^4x^2 + (2a^2b - a^5)x$, a square, $a \neq 0$, $b = 0$, $-\sqrt{2}a^3$, $-a^3$, $-(\sqrt{2} + 1)a^3$, $q = 9$

ix) $x^6 + ax^5 - a^4x^2$, $q = 27$

x) $x^6 + ax^5 + a^3x^3 - a^4x^2 - a^5x$, $q = 9$

xi) $x^6 + ax^5 + ba^3x^3 - ba^4x^2 + \sqrt{2}a^5x$, $b = \pm(1 - \sqrt{2})$ and $q = 9$

xii) $x^6 + ax^5 - a^3x^3 + a^4x^2 + (\sqrt{2} - 1)a^5x$, $q = 9$.

Proof. See Dickson (1897). ∎

Theorem 3.15.

a) The orthomorphism polynomials of degree less than 6 are translates of the following orthomorphism polynomials:

i) ax, $a \neq 0, 1$, for all $q > 2$

ii) $ax^4 + bx^2 + cx$, $a \neq 0$, q even, $q > 2$, and $ax^4 + bx^2 + cx = x$ and $ax^4 + bx^2 + (c - 1)x = x$ only if $x = 0$

iii) $\pm x^4 + 4x$, $q = 7$

iv) $5a^2x^5 + 5ax^3 + 2x$ and $8a^2x^5 + 8ax^3 - x$, a not a square, $q = 13$

v) $\pm\sqrt{2}x^5 \pm x$ and $\pm\sqrt{2}x^5$, $q = 9$

vi) $ax^P + bx$, $a \neq 0$, $q = p^r$, $p = 3$ or 5, and $ax^P + bx = x$ and $ax^P + (b-1)x = x$ only if $x = 0$

b)   The orthomorphism polynomials of degree 6 for q relatively prime to 6 are translates of the following orthomorphism polynomials:

i) $a(x^6 + 2x)$, $a = 2, 3, 5$, $q = 11$

ii) $a(x^6 - 2x)$, $a = 6, 8, 9$, $q = 11$

iii) $a(x^6 + 4x)$, $a = 2, 6, 7$, $q = 11$

iv) $a(x^6 - 4x)$, $a = 4, 5, 9$, $q = 11$

Proof. See Niederreiter and Robinson (1982). ∎

Note that the orthomorphisms listed in Theorem 3.15 are all cyclotomic orthomorphisms or automorphisms. All orthomorphisms in Theorem 3.15 b) are quadratic orthomorphisms, as are iii) and v) of Theorem 3.15 a). The orthomorphisms in Theorem 3.15 a) iv) are in $\mathfrak{C}_6(13)$ and the orthomorphisms of Theorem 3.15 a) i), ii), and vi) are automorphisms. Automorphisms that are also orthomorphisms will be studied in Chapter 4.

Section 4: Cyclotomy and permutation polynomials.

In this section we will characterize orthomorphism polynomials corresponding to cyclotomic orthomorphisms. This will enable us to use Hermite's criterion to obtain more information about the structure of Orth(GF(q)$^+$). First the characterization.

Theorem 3.16. Let θ be an orthomorphism of GF(q)$^+$, $q = ef + 1$, and by abuse of notation let us use $\theta(x)$ to denote the corresponding orthomorphism polynomial whose degree is at most $q - 3$. Then $\theta \in \mathfrak{C}_e(q)$ if and only if $\theta(x) = \sum\limits_{i=0}^{e-1} a_i x^{1+if}$ for some a_i in GF(q).

Proof. Let $\theta(x) = \sum\limits_{i=1}^{q-3} a_i x^i$ be an orthomorphism polynomial of GF(q). Then, by Theorem 3.9, $\theta(x)$ is a cyclotomic polynomial of index e if and only if $\theta(ax) = a\theta(x)$ for all $a \in \mathrm{Gf}(q)$ for which $a^f = 1$. This is true if and only if $a^i = a$ whenever $a_i \neq 0$, i.e. if and only if $i - 1$ is divisible by f. The result follows. ∎

Some results that were referred to in earlier sections but which could not be proved there are proved here. Specifically.

Theorem 3.17. Let q be a prime. If $a, b \mid q - 1$, $a, b \neq q - 1$, and $g, h \in GF(q)$, $g \neq h$, then $T_g[\mathfrak{C}_a] \cap T_h[\mathfrak{C}_b] = \mathfrak{C}_1$.

Proof. Without loss of generality we may assume $h = 0$ and $a, b \neq 1$. Clearly $\mathfrak{C}_1 \subseteq T_g[\mathfrak{C}_a] \cap \mathfrak{C}_b$. Next assume $\theta \in T_g[\mathfrak{C}_a] \cap \mathfrak{C}_b$. Then there exists $\varphi \in \mathfrak{C}_a$ such that $\theta = T_g[\varphi]$. Let $q = af + 1 = bh + 1$. Then by Theorem 3.16

$$\theta(x) = \sum_{i=0}^{a-1} a_i (x + g)^{1 + if} - \sum_{i=0}^{a-1} a_i g^{1 + if} = \sum_{j=0}^{b-1} b_j x^{1 + jh} \text{ for some } a_0, \ldots, a_{a-1},$$

$b_0, \ldots, b_{b-1} \in GF(q)$. Let $I = \sup \{i: a_i \neq 0\}$. If $I \neq 0$ then the coefficients of $x^{1 + If}$ and x^{If} are a_I and $a_I(1 + If)g$ respectively, both of which are non-zero since q is a prime. But comparing coefficients, we see that we cannot have consecutive powers of x with non-zero coefficients. Hence $I = 0$ and $\theta \in \mathfrak{C}_1$. ∎

The result of Theorem 3.17 does not hold when q is not a prime. As an example, let $q = 25$ and let $\theta(x) = Ax^5 + Bx$. Then for appropriate choices of $A, B, A \neq 0$, θ is an orthomorphism of $GF(25)^+$ and $\theta \in (T_g[\mathfrak{C}_6] \cap T_h[\mathfrak{C}_6]) - \mathfrak{C}_1$ for all $g, h \in GF(25)$. For $a = b = 2$, however, Theorem 3.17 still holds when q is not a prime. This and information on adjacencies between elements of $T_g[\mathfrak{C}_2]$ and $T_h[\mathfrak{C}_2]$ is established in the following theorem.

Theorem 3.18.

i) Let q be odd and let $g, h \in GF(q)$, $g \neq h$. Then $T_g[\mathfrak{C}_2] \cap T_h[\mathfrak{C}_2] = \mathfrak{C}_1$.

ii) Let q be odd, $q > 5$, and let $g, h \in GF(q)$, $g \neq h$. If $\theta, \phi \in \mathfrak{C}_2 - \mathfrak{C}_1$ then $T_g[\theta]$ cannot be adjacent to $T_h[\phi]$.

Proof. i) The result is clear if $q = 3$ or 5 as then $\mathfrak{C}_{q-1} = \mathfrak{C}_1$. If $q > 5$ then $\mathfrak{C}_1 \subseteq T_g[\mathfrak{C}_2] \cap T_h[\mathfrak{C}_2]$. Without loss of generality we will assume that $h = 0$. Let $\theta \in T_g[\mathfrak{C}_2] \cap \mathfrak{C}_2$. Then there exists $\phi \in \mathfrak{C}_2$ such that $\theta = T_g[\phi]$. By Theorem 3.16, $\theta(x) = Ax^{(q+1)/2} + Bx$ and $\phi(x) = Cx^{(q+1)/2} + Dx$. Thus $A(x + g)^{(q+1)/2} + B(x + g) - Ag^{(q+1)/2} - Bg = Cx^{(q+1)/2} + Dx$ and comparing coefficients yields $A = C = 0$ and so $\theta \in \mathfrak{C}_1$.

ii) Without loss of generality we may assume that $h = 0$. Let $\theta, \phi \in \mathfrak{C}_2$, $T_g[\theta] \sim \phi$. By Theorem 3.16, $\theta(x) = Ax^{(q+1)/2} + Bx$ and $\phi(x) = Cx^{(q+1)/2} + Dx$ and by Theorem 3.12 the coefficient of x^{q-1} in the reduction of $(T_g[\theta](x) - \phi(x))^2$ modulo $x^q - x$ must be zero. Hence the coefficient of x^{q-1} in the reduction of $T_g[\theta](x)\phi(x)$ modulo $x^q - x$ must be zero. But this coefficient is $ACg^2(q^2 - 1)/8 \neq 0$, from which the result follows. ∎

Niederreiter and Robinson (1982) constructed a class of permutation polynomials as follows.

Lemma 3.1. If $q = ef + 1$, $e \geq 2$, then $x^{f+1} + bx$ is a permutation polynomial of $GF(q)$ if and only if the following conditions hold:

i) $(-b)^e \neq 1$.

ii) $((b + \omega^i)(b + \omega^j)^{-1})^f \neq \omega^{j-i}$ for all $0 \leq i < j < e$, where ω is a fixed primitive eth. root of unity in $GF(q)$.

Proof. Let ω be a primitive eth root of unity in $GF(q)$, $q = ef + 1$, and let $h(x) = x^{f+1} + bx$.

If $h(x)$ is a permutation polynomial then condition i) holds as otherwise $-b = a^f$ for some $a \neq 0$ and so $h(0) = h(a) = 0$. If condition ii) does not hold then for some i, j, $0 \leq i < j < e$, $((b + \omega^i)(b + \omega^j)^{-1})^f = \omega^{j-i}$, i.e. $(b + \omega^i)(b + \omega^j)^{-1} = g^{j-i}$, for some primitive element g of $GF(q)$ satisfying $g^f = \omega$, and so $h(g^j) = h(g^i)$, a contradiction.

Next suppose that conditions i) and ii) hold. If $h(x) = 0$ then either $x = 0$ or $-b = x^f$, in which case $(-b)^e = 1$. Thus $h(x) = 0$ if and only if $x = 0$. If $h(a) = h(c)$ for $a, c \neq 0$ then for some i, j, $a^f = \omega^i$ and $c^f = \omega^j$. But then $(\omega^i + b)a = (\omega^j + b)c$ from which it follows that $((b + \omega^i)(b + \omega^j)^{-1})^f = \omega^{j-i}$. Thus $i = j$ and so $a = c$ and $h(x)$ is a permutation polynomial. ∎

From this Niederreiter and Robinson (1982) constructed complete mapping polynomials. In the following theorem we give their construction in orthomorphism form and determine adjacencies within this class of orthomorphism polynomials.

Theorem 3.19. Let $q \equiv ef + 1$, $e \geq 2$.

a)   If $f(x) = ax^{f+1} + bx$, $a \neq 0$, then $f \in \mathfrak{C}_e(q)$ if and only if the following conditions hold for $c = b/a$, $(b - 1)/a$:

   i) $(-c)^e \neq 1$.

   ii) $((c + \omega^i)(c + \omega^j)^{-1})^f \neq \omega^{j-i}$ for all $0 \leq i < j < n$, where ω is a fixed primitive eth. root of unity in $GF(q)$.

b)   If $f(x) = ax^{f+1} + bx$, $a \neq 0$, then $f \sim dx$ if and only if the following conditions hold for $c = (b - d)/a$:

   i) $(-c)^e \neq 1$.

   ii) $((c + \omega^i)(c + \omega^j)^{-1})^f \neq \omega^{j-i}$ for all $0 \leq i < j < n$, where ω is a fixed primitive eth. root of unity in $GF(q)$.

c)   If $f(x) = ax^{f+1} + bx$ and $f'(x) = a'x^{f+1} + b'x$, $a, a' \neq 0$, then $f \sim f'$ if and only if $a = a'$ and $b \neq b'$ or the following conditions hold for $c = (b - b')/(a - a')$:

   i) $(-c)^e \neq 1$.

   ii) $((c + \omega^i)(c + \omega^j)^{-1})^f \neq \omega^{j-i}$ for all $0 \leq i < j < n$, where ω is a fixed primitive eth. root of unity in $GF(q)$.

Proof. This is a consequence of Lemma 3.1, the fact that if f is a normalized permutation polynomial then $aT_b[f]$ will be an orthomorphism polynomial if and only if $f \sim [a^{-1}]$. ∎

Chapter 4, Automorphisms and translation nets

Section 1.   Introduction

A *translation net* N is a net with a collineation group G that acts regularly on the points and preserves parallel classes. Let N have order n and degree r. Then G will have order n^2 and without loss of generality we may assume the points of N to be the elements of G. Let U be the set of lines through e. The lines of N are $\{gV: g \in G, V \in U\}$, $|U| = r$, and $V \cap V' = \{e\}$ for all distinct $V, V' \in U$. For each line V in U, G acts, by left multiplication, transitively on the n lines of the parallel class containing V. The stabilizer of V is a subgroup of G of order n, which by the regularity of G must be V. Thus each element of U is a subgroup of G.

A *partial congruence partition* of a group G of order n^2 is a set U of r subgroups of G of order n, such that $V \cap V' = \{e\}$ for all distinct $V, V' \in U$. The elements of U are called the *components* of the partial congruence partition, and r is called its *degree*. We have just shown that partial congruence partitions can be constructed from translation nets, and reversing this construction we see that translation nets and partial congruence partitions are equivalent concepts. This equivalence was first established by Sprague (1982).

We are particularly interested in partial congruence partitions in which at least two elements of U are normal subgroups of G. The corresponding translation nets are called *splitting translation nets*. Sprague (1982) showed that, for such partial congruence partitions, any two elements of U will be isomorphic, and $G \cong H \times H$, where $H \times \{e\}$ and $\{e\} \times H$ are elements of U. Let N be the corresponding splitting translation net. Then the mappings $\Pi = \{\pi_h: G \to G, h \in H\}$ defined by $\pi_h((a, b)) = (ha, b)$ form a collineation group of N, isomorphic to H. Now Π fixes all lines of the parallel class containing $H \times \{e\}$, acts sharply transitively on the points of any line in this class, and fixes all parallel classes. Thus, by Theorem 1.12 a), N can be constructed from a clique, $\{\theta_1, \dots, \theta_r\}$ say, of $\mathrm{Orth}(H)$. The set of mappings $\Sigma = \{\sigma_h: G \to G, h \in H\}$ defined by $\sigma_h((a, b)) = (a, hb)$ is also a collineation group of N, that fixes all lines of the parallel class containing $\{e\} \times H$, acts sharply transitively on the points of any line in this class, and fixes all parallel classes. From this we can prove that $T_h[\theta_i] = \theta_i$, for all $h \in H$, and $i = 1, \dots, r$. Hence $\theta_i \in \mathrm{Aut}(H)$ for all i.

Thus all splitting translation nets can be constructed from orthomorphisms that are also automorphisms. The corresponding construction of mutually orthogonal Latin squares from group automorphisms dates back to Mann (1942). In this chapter we will study $\mathcal{Q}(G) = \mathrm{Orth}(G) \cap \mathrm{Aut}(G)$, and a closely related orthomorphism graph $\mathcal{Q}^-(G)$. We will omit the study of projective planes constructed from cliques of $\mathcal{Q}(G)$, i.e. translation planes (see Dembowski (1968)), and translation transversal designs (see Hachenberger and Jungnickel (1990)), restricting ourselves to determining some elementary facts about the structure of $\alpha(G) \cup \alpha^-(G)$, bounds for $\omega(\alpha(G) \cup \alpha^-(G))$, and, when G is abelian, exact values of $\omega(\alpha(G))$.

Section 2.   Orthomorphisms from automorphisms

We now turn our attention to the construction of orthomorphisms and complete mappings from automorphisms of groups. We say that $\alpha \in \text{Aut}(G)$ is *fixed point free* if $\alpha(x) = x$ implies that $x = e$. It turns out that the set of automorphisms that are also orthomorphisms is precisely the set of fixed point free automorphisms. Let us define α^- by $\alpha^-(x) = \alpha(x)^{-1}$. In the following theorem we characterize those automorphisms of a group that are orthomorphisms of the group, and those automorphisms of a group that are complete mappings of the group.

Theorem 4.1. Let $\alpha \in \text{Aut}(G)$.

i) α is an orthomorphism of G if and only if α is fixed point free.

ii) α is a complete mapping of G if and only if $\alpha(x) = y^{-1} x^{-1} y$, for some y, implies that $x = e$, if and only if α^- is an orthomorphism of G.

Proof.

i)   Clearly for an automorphism to be an orthomorphism it must be fixed point free. Next, suppose that $\alpha \in \text{Aut}(G)$ is fixed point free. Then $\alpha \in \text{Orth}(G)$ if and only if the mapping $x \to x^{-1}\alpha(x)$ is a bijection. This is the case as $x^{-1}\alpha(x) = y^{-1}\alpha(y)$ if and only if $\alpha(xy^{-1}) = xy^{-1}$ if and only if $x = y$.

ii)   Note that $x\alpha(x) = y\alpha(y)$ if and only if $\alpha(xy^{-1}) = x^{-1}y$ if and only if $\alpha(z) = y^{-1}z^{-1}y$, where $z = xy^{-1}$. Thus $\alpha \in \text{Aut}(G)$ is a complete mapping if and only if $\alpha(x) = y^{-1} x^{-1} y$, for some y, implies that $x = e$. Now α^- is an orthomorphism if and only if $x^{-1}\alpha^-(x) = y^{-1}\alpha^-(y)$ implies that $x = y$ if and only if $x^{-1}\alpha(x^{-1}) = y^{-1}\alpha(y^{-1})$ implies that $x = y$ if and only if α is a complete mapping. ∎

It should be noted that, in Theorem 4.1, we cannot replace the condition that $\alpha(x) = y^{-1} x^{-1} y$, for some y, implies that $x = e$, by the condition that $\alpha(x) = x^{-1}$ implies that $x = e$ on account of the following example.

Let $G = S_3$, the symmetric group on 3 symbols, which can be defined as $<a, b: a^3 = b^2 = e, ab = ba^2>$ and consider the following table.

x	$\alpha(x)$	x^{-1}
e	e	e
a	a	a^2
a^2	a^2	a
b	a^2b	b
ab	b	ab
a^2b	ab	a^2b

We see that α is an automorphism and $\alpha(x) = x^{-1}$ implies that $x = e$ but $\alpha(a) = b^{-1}a^{-1}b$. In fact this group admits no orthomorphisms or complete mappings as its Sylow 2-subgroup is cyclic.

Let us define $\mathcal{Q}(G) = \text{Aut}(G) \cap \text{Orth}(G)$ and $\mathcal{Q}^-(G) = \{\alpha^-: \alpha \in \text{Aut}(G)\} \cap \text{Orth}(G)$. In Theorem 4.1 we have characterized the elements of $\mathcal{Q}(G)$ and $\mathcal{Q}^-(G)$. In the next theorem we characterize adjacencies in $\mathcal{Q}(G) \cup \mathcal{Q}^-(G)$.

Theorem 4.2. Let $\alpha, \beta \in \text{Aut}(G)$.

i) $\alpha \sim \beta$ if and only if $\beta\alpha^{-1}$ is fixed point free.

ii) $\alpha^- \sim \beta^-$ if and only if $\alpha \sim \beta$.

iii) $\alpha^- \sim \beta$ if and only if $\alpha\beta^{-1}$ is a complete mapping.

Proof.

i)   This is an immediate consequence of Corollary 1.1 and Theorem 4.1 i).

ii) This follows from the fact that $\alpha^-(x)^{-1}\beta^-(x) = \alpha(x^{-1})^{-1}\beta(x^{-1})$.

iii) This follows from the fact that $\beta(x)^{-1}\alpha^-(x) = \beta(x^{-1})\alpha(x^{-1}) = \beta(x^{-1})\alpha\beta^{-1}(\beta(x^{-1}))$. ∎

Corollary 4.1. Let $\mathcal{H}$ be a fixed point free automorphism group of G (i.e. every non-identity element of $\mathcal{H}$ is a fixed point free automorphism of G). Then $\mathcal{H}$ - {identity} is a clique of $\text{Orth}(G)$.

Proof. Let $\alpha, \beta \in \mathcal{H}$ - {identity}, $\alpha \neq \beta$. Then $\alpha\beta^{-1} \in \mathcal{H}$ - {identity} and so is an orthomorphism and so $\alpha \sim \beta$. ∎

Section 3.   Bounds on $\omega(\mathcal{Q}(G) \cup \mathcal{Q}^-(G))$

In this section we will derive upper bounds on the clique number of $\mathcal{Q}(G) \cup \mathcal{Q}^-(G)$. These bounds can be found in Evans (1989d). For $\mathcal{Q}(G)$ many of these bounds are special cases of bounds obtained by Jungnickel (1989a) for more general translation nets.

Let τ_g denote the mapping defined by $\tau_g(x) = gxg^{-1}$, i.e. τ_g is an inner automorphism of G.

Lemma 4.1. Let α and $\beta \in \mathrm{Aut}(G)$ and let $g, h \in G$. Then the following hold.

a) $\tau_g\alpha \in \mathrm{Orth}(G)$ if and only if $\alpha \in \mathrm{Orth}(G)$.

b) $\tau_g\alpha^{\cdot} \in \mathrm{Orth}(G)$ if and only if $\alpha^{\cdot} \in \mathrm{Orth}(G)$.

c) $\tau_g\alpha \sim \tau_h\beta$ if and only if $\alpha \sim \beta$.

d) $\tau_g\alpha^{\cdot} \sim \tau_h\beta^{\cdot}$ if and only if $\alpha^{\cdot} \sim \beta^{\cdot}$.

e) $\tau_g\alpha^{\cdot} \sim \tau_h\beta$ if and only if $\alpha^{\cdot} \sim \beta$.

f) α is not adjacent to $\tau_g\alpha$.

g) $\alpha^{\cdot}$ is not adjacent to $\tau_g\alpha^{\cdot}$.

Proof.

a)   If $\alpha \in \mathrm{Orth}(G)$ then α is a fixed point free automorphism of G and $\tau_g\alpha(x) = x$ implies $\alpha(x) = g^{-1}xg$, which implies that $x = e$ by Huppert (1967, Satz 8.9), which implies that $\tau_g\alpha$ is a fixed point free automorphism of G, which implies that $\tau_g\alpha \in \mathrm{Orth}(G)$. Similarly if $\tau_g\alpha \in \mathrm{Orth}(G)$ then $\tau_{g^{-1}}\tau_g\alpha = \alpha \in \mathrm{Orth}(G)$.

b)   If $\alpha^{\cdot} \in \mathrm{Orth}(G)$ then $\alpha(x) = y^{-1}x^{-1}y$ implies $x = e$ and $\tau_g\alpha(x) = y^{-1}x^{-1}y$ implies implies $\alpha(x) = (yg)^{-1}x^{-1}(yg)$, which implies that $x = e$. Thus $(\tau_g\alpha)^{\cdot} = \tau_g\alpha^{\cdot} \in \mathrm{Orth}(G)$. Similarly if $\tau_g\alpha^{\cdot} \in \mathrm{Orth}(G)$ then $\tau_{g^{-1}}\tau_g\alpha^{\cdot} = \alpha^{\cdot} \in \mathrm{Orth}(G)$.

c) $\tau_g\alpha \sim \tau_h\beta$ if and only if $\tau_h\beta(\tau_g\alpha)^{-1} = \tau_h(\beta\alpha^{-1})\tau_{g^{-1}}$ is a fixed point free automorphism of G. But $\tau_h(\beta\alpha^{-1})\tau_{g^{-1}}(x) = x$ if and only if $\beta\alpha^{-1}(g^{-1}xg) =$ $(h^{-1}g)(g^{-1}xg)(h^{-1}g)^{-1}$. Thus, by Huppert (1967, Satz 8.9), $\tau_h(\beta\alpha^{-1})\tau_{g^{-1}}$ is a fixed point free automorphism of G if and only if $\beta\alpha^{-1}$ is a fixed point free automorphism of G.

d)   This is an immediate consequence of c) and Theorem 4.2 ii).

e)   By Theorem 4.2 iii), $\tau_g\alpha^{\cdot} \sim \tau_h\beta$ if and only if $\tau_g\alpha^{\cdot}(\tau_h\beta)^{-1}$ is a complete mapping of G. By Theorem 4.1 ii), $\tau_g\alpha^{\cdot}(\tau_h\beta)^{-1}$ is a complete mapping of G if and only if $\tau_g\alpha^{\cdot}\beta^{-1}\tau_{h^{-1}}(x) = y^{-1}x^{-1}y$ implies $x = e$, if and only if $\alpha^{\cdot}\beta^{-1}(h^{-1}xh) =$ $(h^{-1}yg)^{-1}h^{-1}x^{-1}h(h^{-1}yg)$ implies $x = e$, if and only if $\alpha\beta^{-1}$ is a complete mapping of G, if and only if $\alpha^{\cdot} \sim \beta$.

f) $\alpha \sim \tau_g\alpha$ if and only if $\tau_g\alpha\alpha^{-1} = \tau_g$ is a fixed point free automorphism of G, which is impossible by Huppert (1967, Satz 8.9).

g)   This is an immediate consequence of f) and Theorem 4.2 ii). ∎

Theorem 4.3. Let H be a non-trivial subgroup of G and let $\mathfrak{H} = \{gHg^{-1}: g \in G\}$. Let $A_{\mathfrak{H}} = \{\alpha: \alpha \text{ or } \alpha^{\cdot} \in \mathrm{Aut}(G), \alpha \text{ fixes } \mathfrak{H}\}$ and let $A_H = \{\alpha: \alpha \text{ or } \alpha^{\cdot} \in \mathrm{Aut}(G), \alpha \text{ fixes } H\}$. Then $\omega(A_{\mathfrak{H}} \cap \mathrm{Orth}(G)) \le \omega(A_H \cap \mathrm{Orth}(H))$.

Proof. If $\alpha_1, \ldots, \alpha_n$ is a clique of $A_{\mathfrak{H}} \cap \mathrm{Orth}(G)$ then $\alpha_i(H) = H_i = g_i^{-1}Hg_i$, for some $g_i \in G$. Now $\beta_i = \tau_{g_i}\alpha_i$, fixes H and $\beta_1|_H, \ldots, \beta_n|_H$ is a clique of $A_H \cap \mathrm{Orth}(H)$. ∎

From Theorem 4.3 we can immediately derive the following bounds on $\omega(\mathcal{C}(G) \cup \mathcal{C}^-(G))$.

Corollary 4.2. $\omega(\mathcal{C}(G) \cup \mathcal{C}^-(G)) \leq$
Min$\{\omega(\mathcal{C}(H) \cup \mathcal{C}^-(H))$: $p \mid |G|$, H a Sylow p-subgroup of $G\}$.

Corollary 4.3. If $Z(G)$ is non-trivial then
$\omega(\mathcal{C}(G) \cup \mathcal{C}^-(G)) \leq \omega(\mathcal{C}(Z(G)) \cup \mathcal{C}^-(Z(G)))$.

Corollary 4.4. If $Z(G)$ is non-trivial then $\omega(\mathcal{C}(G) \cup \mathcal{C}^-(G)) \leq$
Min$\{\omega(\mathcal{C}(H) \cup \mathcal{C}^-(H))$: $p \mid |Z(G)|$, H a Sylow p-subgroup of $Z(G)\}$.

Corollary 4.5. $\omega(\mathcal{C}(G) \cup \mathcal{C}^-(G)) \leq$
Min$\{\omega(\mathcal{C}(Z(H)) \cup \mathcal{C}^-(Z(H)))$: $p \mid |G|$, H a Sylow p-subgroup of $G\}$.

Corollary 4.6. Let H be a non-trivial characteristic subgroup of G. Then
$\omega(\mathcal{C}(G) \cup \mathcal{C}^-(G)) \leq \omega(\mathcal{C}(H) \cup \mathcal{C}^-(H))$.

Corollary 4.7. $\omega(\mathcal{C}(G) \cup \mathcal{C}^-(G)) = |G| - 2$ if and only if G is an elementary abelian group.

Section 4.   The abelian case

If G is abelian then $\mathcal{C}^-(G) = \mathcal{C}(G)$, and so $\omega(\mathcal{C}(G) \cup \mathcal{C}^-(G)) = \omega(\mathcal{C}(G))$. The exact value of $\omega(\mathcal{C}(G))$, for G abelian, was determined by Bailey and Jungnickel (1990). In this section we will derive Bailey and Jungnickel's result.

Theorem 4.4. (Huppert, p.501, Satz 8.10) Let α be a fixed point free automorphism of G and let H be a normal subgroup of G with $\alpha(H) = H$. Then α induces a fixed point free automorphism on G/H.
Proof. The mapping α': $gH \to \alpha(g)H$ is clearly an automorphism of G/H. Suppose $\alpha(g)H = gH$. Then $g^{-1}\alpha(g)H = H$ and so $g^{-1}\alpha(g) \in H$. But then $g \in H$, as $\alpha \in$ Orth(G) and $\alpha|_H \in$ Orth(H), and so α' is a fixed point free automorphism of G/H. ∎

Corollary 4.8. Let α, $\beta \in \mathcal{C}(G)$, $\alpha \sim \beta$, and let H be a normal subgroup of G with $\alpha(H) = \beta(H) = H$. If α' and β' are the corresponding fixed point free automorphisms induced on G/H then $\alpha' \sim \beta'$.
Proof. The assumption that $\alpha \sim \beta$ implies that $\alpha\beta^{-1} \in \mathcal{C}(G)$ and so $\alpha\beta^{-1}$ induces a fixed

point free automorphism on G/H, which we shall denote γ'. Now $\gamma'(gH) = \alpha\beta^{-1}(g)H = \alpha'(\beta^{-1}(g)H) = \alpha'(\beta^{-1})'(gH)$ and so $\gamma' = \alpha'(\beta^{-1})'$. But $(\beta^{-1})' = \beta'^{-1}$ as $\beta'(\beta^{-1})'(gH) = \beta\beta^{-1}(g)H = gH$. Hence $\gamma' = \alpha'\beta'^{-1}$ and so $\alpha' \sim \beta'$. ∎

Corollary 4.9. If H char G then $\omega(\mathcal{C}(G)) \leq \text{Min} \{\omega(\mathcal{C}(H)), \omega(\mathcal{C}(G/H))\}$.

Corollary 4.10.

i) $\omega(\mathcal{C}(G \times H)) \geq \text{Min} \{\omega(\mathcal{C}(H)), \omega(\mathcal{C}(G))\}$.

ii)   If $|G|$ and $|H|$ are relatively prime then $\omega(\mathcal{C}(G \times H)) = \text{Min} \{\omega(\mathcal{C}(H)), \omega(\mathcal{C}(G))\}$.

Proof. This follows from Corollary 4.9 and the construction of Theorem 1.3. ∎

Corollary 4.11. If G is nilpotent then
$\omega(\mathcal{C}(G)) = \text{Min}\{\omega(\mathcal{C}(H)): p \mid |G|, H \text{ a Sylow } p\text{-subgroup of } G\}$.

Thus, to determine $\omega(\mathcal{C}(G))$ for abelian groups, we need only determine $\omega(\mathcal{C}(G))$ for abelian p-groups. Let us use nZ_q to denote the direct product of n copies of Z_q. Then any abelian p-group, p a prime, can be uniquely written as $m_k Z_{q_k} \times \ldots \times m_1 Z_{q_1}$, where $q_1 < \ldots < q_k$ are powers of p. Thus a first step toward determining $\omega(\mathcal{C}(G))$, for G abelian, is to determine $\omega(\mathcal{C}(m_i Z_{q_i}))$, for q_i a prime power.

Lemma 4.2. Jungnickel (1989b). If q is a power of a prime p then $\omega(\mathcal{C}(mZ_q)) = p^m - 2$.

Proof. Let $G = mZ_q$. Then $H = \{g: g \in G, g^P = e\}$ is a characteristic subgroup of G. As $H \cong mZ_p$, by Corollary 4.9, $\omega(\mathcal{C}(G)) \leq p^m - 2$. To show equality, let $\theta_1, \ldots, \theta_r$ be an r-clique of $\mathcal{C}(mZ_p)$. The action of an automorphism on a chosen set of minimal generators X of mZ_p (respectively mZ_q) uniquely determines the automorphism. Using this action we can construct a unique automorphism θ'_i of mZ_q, for each i, whose action on a chosen set of generators Y of mZ_q is formally the same as the action of θ_i on X. $\theta'_1, \ldots, \theta'_r$ will form an r-clique of $\mathcal{C}(mZ_q)$. ∎

We are now in a position to determine $\omega(\mathcal{C}(G))$, for G an abelian p-group.

Theorem 4.5. Bailey and Jungnickel (1990). Let $m = \text{Min} \{m_i: i = 1, \ldots, k\}$ and $G = m_k Z_{q_k} \times \ldots \times m_1 Z_{q_1}$, where $q_1 < \ldots < q_k$ are powers of a prime p. Then
$\omega(\mathcal{C}(G)) = p^m - 2$.

Proof. By Corollary 4.10 and Lemma 4.2, $\omega(\mathcal{C}(G)) \geq \text{Min} \{\omega(\mathcal{C}(m_k Z_{q_k})), \ldots, \omega(\mathcal{C}(m_1 Z_{q_1}))\} = p^m - 2$. Next set $H_1 = \{g: g \in G, g^P = e\}$,

$H_i = \{h: h \in H_1, g^{q_{i-1}} = h \text{ for some } g \in G\}$ for $i = 2, \ldots, k$, and $H_{k+1} = \{e\}$.

Then $H_{i+1} \subseteq H_i$, H_i char G, and $|H_i| = p^{m_k + \cdots + m_i}$, for $i = 1, \ldots, k$. Thus $|H_i/H_{i+1}| = p^{m_i}$. Hence, by Corollary 4.9, $\omega(\mathcal{C}(G)) \leq p^m - 2$ and so $\omega(\mathcal{C}(G)) = p^m - 2$. ∎

Chapter 5, Further results

Section 1:   Powers.

In this section we will study orthomorphism graphs induced by mappings of the form $\phi_r \colon G \to G$, $\phi_r(x) = x^r$. $\mathcal{P}(G)$ will denote the set $\{\phi_r \colon \phi_r \in \text{Orth}(G)\}$ and the induced orthomorphism graph. We will see that the structure of $\mathcal{P}(G)$ depends not on the structure of G, but on the order of G. It is very easy to determine which maps ϕ_r belong to $\mathcal{P}(G)$ and to determine adjacencies in $\mathcal{P}(G)$. However the structure of $\mathcal{P}(G)$ can be quite complex. It turns out that any finite graph can be realized as an induced subgraph of $\mathcal{P}(G)$ for some group G. Note that, if G is abelian then $\mathcal{P}(G) \subseteq \mathcal{Q}(G)$.

Theorem 5.1. If G is a group of order n then the following hold.

i) ϕ_r is a permutation if and only if $(r, n) = 1$.

ii) ϕ_r is an orthomorphism if and only if $(r, n) = (r - 1, n) = 1$.

iii) ϕ_r is a complete mapping if and only if $(r, n) = (r + 1, n) = 1$.

iv) $\phi_r \sim \phi_s$ if and only if $(s - r, n) = 1$.

Proof. i) If $(r, n) \neq 1$ then let p be a prime divisor of both r and n. Let $g \in G$ have order p. Then $g^r = e$ and so ϕ_r cannot be a permutation. On the other hand, if $(r, n) = 1$ then there exists r' satisfying $r'r \equiv 1$ modulo n. Thus $\phi_r(x^{r'}) = x$ and so ϕ_r is onto and hence a permutation. ii), iii), and iv) follow immediately from i). ∎

From this we can determine $\omega(\mathcal{P}(G))$ and the size of all maximal cliques of $\mathcal{P}(G)$.

Corollary 5.1. If G is a group of order n then $\omega(\mathcal{P}(G)) = p - 2$, where p is the smallest prime dividing n. Further, any clique of $\mathcal{P}(G)$ can be extended to a $(p - 2)$ - clique of $\mathcal{P}(G)$.

Proof. $\phi_2, \dots, \phi_{p-1}$ is a $(p - 2)$ - clique of $\mathcal{P}(G)$ and, by the pigeonhole principle, any set $\{0, 1, a_2, \dots, a_p\}$ of integers must contain a pair of integers that are congruent modulo p, and so no larger clique of $\mathcal{P}(G)$ exists.

To prove the extendibility of cliques it is sufficient to prove that any r - clique of $\mathcal{P}(G)$ can be extended to an $(r + 1)$ - clique of $\mathcal{P}(G)$ if $r < p - 2$. Let $\phi_{a_1}, \dots, \phi_{a_r}$ be an r - clique of $\mathcal{P}(G)$, $r < p - 2$, and let $p = p_1 < \dots < p_m$ be the prime divisors of n. Now for each j there exists c_j such that $c_j \not\equiv 0, 1$ modulo p_j and $c_j \not\equiv a_i$ modulo p_j for $i = 1, \dots$, r. By the chinese remainder theorem there exists $a \equiv c_j$ modulo p_j for all j, $1 < a < n$. Clearly $\phi_a \in \mathcal{P}(G)$ and $\phi_a \sim \phi_{a_i}$ for all i, thus extending our clique. ∎

Theorem 1.22, proving the existence of orthomorphisms of groups of odd order, is a special case of Corollary 5.1, as ϕ_2 is an orthomorphism of any group of odd order. Equivalently ϕ_1 is a complete mapping of any group of odd order. Another consequence of this corollary is that $\mathcal{P}(G)$ is nonempty if and only if $|G|$ is odd.

In Corollary 5.1 we have shown certain cliques to be maximal in $\mathcal{P}(G)$. This does not imply their maximality in Orth(G). In section 4 of this chapter we will establish that for some groups, in particular cyclic groups, maximal cliques of $\mathcal{P}(G)$ are also maximal in Orth(G).

An additional, and immediate, corollary of Theorem 5.1.

Corollary 5.2. If G_1 and G_2 are groups for which $|G_1| = |G_2|$ then $\mathcal{P}(G_1) \cong \mathcal{P}(G_2)$.

The next theorem summarizes the actions of known automorphisms and congruences, of Orth(G), on $\mathcal{P}(G)$.

Theorem 5.2. Let G be a group of order n.

i) $H_\alpha[\phi_r] = \phi_r$ for all $\alpha \in$ Aut(G).

ii) $T_g[\phi_r] = \phi_r$ for all $g \in Z(G)$.

iii) $R[\phi_r] = \phi_{1-r}$.

iv) $I[\phi_r] = \phi_{r'}$, where r' is the unique integer between 1 and $n-1$ that satisfies $r'r \equiv 1$ modulo n.

Proof. These are routine calculations. ∎

Next we will see that the structure of $\mathcal{P}(G)$ is actually fairly complex as any finite graph can be regarded as an induced subgraph of some $\mathcal{P}(G)$. We will first establish a more general result. Let Γ be a graph with vertices $v_1, \dots, v_r$ and let n be a natural number. We say that Γ is representable modulo n if there exist distinct integers $a_1, \dots, a_r$, $0 \le a_i < n$, satisfying $(a_i - a_j, n) = 1$ if and only if v_i is adjacent to v_j. We call $\{a_1, \dots, a_r\}$ a *representation* of Γ modulo n and we call n the *order* of the representation. If $\{a_1, \dots, a_r\}$ is a representation of Γ modulo n then so is $\{ba_1 + c, \dots, ba_r + c\}$, where $(b, n) = 1$ and addition and multiplication are performed modulo n.

Erdös and Evans (1989) proved that any graph is representable modulo some positive integer. The proof will require the following lemma.

Lemma 5.1. For any positive integer m there exist distinct primes $p_1, \dots, p_m$ such that for all pairs A, B of disjoint, nonempty subsets of $\{p_1, \dots, p_m\}$,

$$(\Pi\{p_i \colon p_i \in A\} - \Pi\{p_j \colon p_j \in B\}, p_1 p_2 \cdots p_m) = 1.$$

Proof. Suppose that we have chosen a prime $p_1 > 3^m$ and further suppose that we have chosen primes $p_1 < \dots < p_s$, $s < m$, to satisfy the conditions of the lemma. We note that in choosing the next prime sufficiently large the only restriction is that p_{s+1} cannot be congruent to a/b modulo p_i, $i = 1, \dots, s$, where a and b are square free products using the primes

$p_1, \ldots, p_s$, and $a \neq 1$. Thus for each i, $i = 1, \ldots, s$, there are at most $3^{s-1} < 3^m < p_1 < p_i$ residue classes to be avoided. Thus for some nonzero residue class modulo $p_1 \ldots p_s$ any choice of a prime in this residue class would be a valid choice for the next prime. A theorem of Dirichlet guarantees the existence of an infinite number of primes in such a residue class, enabling us to extend our selection. The proof follows by induction. ∎

We are now in a position to prove the result of Erdös and Evans (1989) on the representability of graphs.

Theorem 5.3. Any finite graph can be represented modulo some positive integer.

Proof. Let Γ be a graph with vertices $v_1, \ldots, v_r$. Form a new graph Γ' by adjoining an isolated vertex v_0 to Γ. Let $e_1, \ldots, e_m$ be the edges of the complement of Γ' and let $p_1, \ldots, p_m$ be primes satisfying the conditions of Lemma 5.1. Then for $i = 1, \ldots, r$ set $a_i = \Pi \{p_j \colon e_j \text{ incident with } v_i \text{ in the complement of } \Gamma'\}$ and set $n = p_1 p_2 \ldots p_m$. Then $\{a_1, \ldots, a_r\}$ is a representation of Γ modulo n. ∎

Corollary 5.3. i) Any finite graph can be realized as an induced subgraph of $\mathcal{P}(G)$ for some finite group G.

ii) If G_1 and G_2 are two groups of the same order then a given finite graph can be realized as an induced subgraph of $\mathcal{P}(G_1)$ if and only if it can be realized as an induced subgraph of $\mathcal{P}(G_2)$.

Proof. i) Let Γ be a finite graph with vertices $v_1, \ldots, v_r$ and adjoin to Γ vertices u and v adjacent to each of $v_1, \ldots, v_r$ and to each other. Let $\{0, 1, a_1, \ldots, a_r\}$ be a representation of $\Gamma \cup \{u, v\}$ modulo n. Then the subgraph of $\mathcal{P}(G)$ induced by the orthomorphisms $\phi_1, \ldots, \phi_r$ is isomorphic to Γ. Thus Γ is realized as an induced subgraph of $\mathcal{P}(G)$.

ii) This is an immediate consequence of Corollary 5.2. ∎

Lindner, Mendelsohn, Mendelsohn, and Wolk (1979) defined an orthogonal Latin square graph to be a graph, all of whose vertices are Latin squares of the same order, adjacency being orthogonality. They proved that any finite graph is realizable as an orthogonal Latin square graph. Their theorem is an immediate consequence of Corollary 5.3.

Corollary 5.4. Any finite graph can be realized as an orthogonal Latin square graph.

Lindner, Mendelsohn, Mendelsohn, and Wolk actually proved that any finite graph can be realized as an orthogonal Latin square graph using idempotent Latin squares. This extra condition is a by-product of the construction used in their proof.

It should be noted that for a graph to be representable modulo n it is necessary though not sufficient for the smallest prime divisor of n to be at least as large as the clique number of the graph. This is in marked contrast to the situation for orthogonal Latin square realizations, in which Lindner, Mendelsohn, Mendelsohn, and Wolk proved that only finitely many Latin square orders are ruled out for a given graph.

Section 2. Strong complete mappings

A *strong complete mapping* is a complete mapping that is also an orthomorphism. The term strong complete mapping was first used by Hsu and Keedwell (1985, Definition 6.3). They gave a construction for strong complete mappings of odd order elementary abelian groups. Anderson (1984) called these strong orthomorphisms and used them in the construction of strong partial starters, and Horton (1990) called these strong permutations and used them in the construction of strong starters. We shall use $\mathcal{S}(G)$ to denote the set of strong complete mappings of G and the corresponding orthomorphism graph. Note that, for $|G|$ odd, the existence of a strong complete mapping of G implies the existence of a 2 - clique of Orth(G), as any strong complete mapping of G will be adjacent to ϕ_{-1}, an orthomorphism of G. In particular, for $|G|$ odd, $\omega(\mathcal{S}(G)) \leq \omega(\mathrm{Orth}(G)) - 1$.

It should also be clear from this discussion that any automorphism or congruence of Orth(G) that fixes ϕ_{-1} must also fix $\mathcal{S}(G)$. In particular $\mathcal{S}(G)$ is fixed by H_α, $\alpha \in \mathrm{Aut}(G)$, T_g, $g \in Z(G)$, and I.

In this section our main concern will be with the existence of strong complete mappings. Our results on the structure of $\mathcal{S}(G)$ will be fairly rudimentary and are included in Examples 5.1, ... , 5.4.

Example 5.1. $\mathcal{P}(G) \cap \mathcal{S}(G) = \{\phi_r: (r, n) = (r - 1, n) = (r + 1, n) = 1\}$. If $|G|$ is odd and p is the smallest prime divisor of $|G|$ then $\omega(\mathcal{P}(G) \cap \mathcal{S}(G)) = p - 3$. In particular $\mathcal{P}(G) \cap \mathcal{S}(G)$ is nonempty if and only if $(6, |G|) = 1$.

Example 5.2. $\mathcal{C}_1(q) \cap \mathcal{S}(\mathrm{GF}(q)^+) = \{[A]: A \neq 0, \pm1\}$. Thus $\mathcal{C}_1(q) \cap \mathcal{S}(\mathrm{GF}(q)^+)$ is nonempty if and only if $q > 3$, in which case $\omega(\mathcal{C}_1(q) \cap \mathcal{S}(\mathrm{GF}(q)^+)) = q - 3$.

Example 5.3. $\mathcal{C}_2(q) \cap \mathcal{S}(\mathrm{GF}(q)^+) = \{[A, B]: AB, (A - 1)(B - 1), \text{ and}$
$(A + 1)(B + 1) \text{ nonzero squares}\}$. The clique number of $\mathcal{C}_2(q) \cap \mathcal{S}(\mathrm{GF}(q)^+)$ is the same as that for $\mathcal{C}_1(q) \cap \mathcal{S}(\mathrm{GF}(q)^+)$.

Example 5.4. If G is abelian then
$\mathcal{A}(G) \cap \mathcal{S}(G) = \{\alpha: \alpha \in \mathrm{Aut}(G), \alpha(x) \neq x \text{ or } x^{-1} \text{ if } x \neq e\}$. Note that if G is abelian then $\mathcal{A}^-(G) = \mathcal{A}(G)$.

We will now prove that a group with a cyclic Sylow 3-subgroup and a homomorphism onto this subgroup cannot admit a strong complete mapping. This was proved in Horton (1990) and independently in Evans (1990). From this we derive known non-existence results for Knut Vic designs.

Theorem 5.4. Let G be a finite group with a nontrivial cyclic Sylow 3-subgroup S and let H be a normal subgroup of G for which $G/H \cong S$. Then G does not admit strong complete mappings.

Proof. Let $|G| = mn$, $(n, m) = 1$, m a power of 3, and let $\phi: G \to Z_m \cong G/H$ be the canonical homomorphism. If G admits a strong complete mapping θ then the following holds.

$$2 \sum_{g \in G} \phi(g)^2 = \sum_{g \in G} \phi(g\,\theta(g))^2 + \sum_{g \in G} \phi(g^{-1}\theta(g))^2 =$$

$$\sum_{g \in G} (\phi\theta(g) + \phi(g))^2 + \sum_{g \in G} (\phi\theta(g) - \phi(g))^2 =$$

$$2 \sum_{g \in G} \phi\theta(g)^2 + 2 \sum_{g \in G} \phi(g)^2 = 4 \sum_{g \in G} \phi(g)^2$$

This implies that $\sum_{g \in G} \phi(g)^2 = 0$. But by direct computation we obtain the following.

$$\sum_{g \in G} \phi(g)^2 \equiv n\,(0^2 + 1^2 + \ldots + (m-1)^2) \equiv n\,(m-1)m\,(2m-1)/6 \not\equiv 0 \pmod{m}$$

A contradiction from which the result follows. ∎

We now turn to the problem of which groups admit strong complete mappings. We already know of two classes of groups that admit strong complete mappings, elementary abelian groups of order at least 4, and groups whose order is not divisible by either 2 or 3. The following existence results were proved in Horton (1990) and independently in Evans (1990). We first show how strong complete mappings of a group can be constructed from strong complete mappings of normal subgroups and the corresponding quotient groups.

Theorem 5.5. Let H be a subgroup of an abelian group G and let G/H and H both admit strong complete mappings. Then G admits a strong complete mapping.

Proof. Let $\theta: H \to H$ and $\phi: G/H \to G/H$ be strong complete mappings of H and G/H, respectively. Let $g_1, \ldots, g_n$ be a set of left coset representatives of H. Define $\alpha: G/H \to G$ by $\alpha(g_i H) = g_i$ and $\beta: G \to G$ by $\beta(g_i h) = \alpha\phi(g_i H)\theta(h)$, $h \in H$. Then β is a strong complete mapping of G. ∎

To prove this result we needed the group G to be abelian. Can we remove this condition? Clearly if G is the direct product of two groups H and K which both admit strong complete mappings then the construction of Theorem 1.3 enables us to construct strong complete mappings of G. We now prove that all noncyclic abelian 2-groups admit strong complete mappings. To do so we need the following lemma.

Lemma 5.2. If $G \cong Z_m \times Z_2$, $m = 2^n$, then G admits a strong complete mapping.

Proof. The reader can easily verify that the orthomorphism of $Z_m \times Z_2$ constructed in the proof of Theorem 1.21 is also a complete mapping of $Z_m \times Z_2$. ∎

Theorem 5.6. Any noncyclic abelian 2-group admits strong complete mappings.

Proof. If G is a noncyclic abelian 2-group then it is an exercise in group theory to show that G admits a subnormal series $\{e\} = K_1 < \ldots < K_s = G$, $K_i/K_{i-1} \cong GF(q_i)^+$, $q_i > 2$, $i < s$, $K_s/K_{s-1} \cong Z_m \otimes Z_2$, $m = 2^n$. Hence Lemma 5.2 implies that G admits a strong complete mapping. ∎

A Knut Vic design of order n is a Latin square of order n in which, for $i = 0, \ldots, n - 1$, the set of cells $(0, i), (1, 1 + i), \ldots, (n - 1, n - 1 + i)$, addition modulo n, and the set of cells $(0, i), (1, i - 1), \ldots, (n - 1, i - n + 1)$, subtraction modulo n, both contain each symbol of the Latin square exactly once. Sets of cells of the form $(0, i), (1, 1 + i), \ldots, (n - 1, n - 1 + i)$, addition modulo n, are called left diagonals and sets of cells of the form $(0, i), (1, i - 1), \ldots, (n - 1, i - n + 1)$, subtraction modulo n, are called right diagonals.

Theorem 5.7. A Knut Vic design of order n exists if and only if Z_n admits a strong complete mapping.

Proof. Let K be a Knut Vic design of order n and let a be an entry of K. Then the mapping θ, defined by $\theta(j) = a - i$ if and only if the ijth. entry is a, is a strong complete mapping of Z_n.

Conversely let θ be a strong complete mapping of Z_n. Then the square with ijth. entry $i + \theta(j)$ is a Knut Vic design of order n. ∎

Note that the Knut Vic design constructed from θ in Theorem 5.7 is in fact the Latin square that we denoted by $L(\theta)$ in Chapter 1, Section 2. The following existence results for Knut Vic designs follow from Theorem 5.7, Theorem 5.4, Example 5.1, and Theorem 1.20.

Corollary 5.5. Hedayat (1977). A Knut Vic design of order n cannot exist if either 2 or 3 divides n.

Corollary 5.6. Hedayat and Federer (1975). Knut Vic designs of order n exist whenever $(6, n) = 1$.

Hedayat and Federer (1975) proved this by first using implicitly the mapping $x \rightarrow -2x$ to construct Knut Vic designs of order p^r, p a prime other than 2 or 3. They then used a Kronecker product construction to show that Knut Vic designs existed for all orders not divisible by 2 or 3.

Two Knut Vic designs are said to be orthogonal if they are orthogonal as Latin squares. We thus obtain the following from Example 5.1.

Theorem 5.8.
a)   Afsarinejad (1987). If p is the smallest prime divisor of n then there exist at least $p - 3$ mutually orthogonal Knut Vic designs of order n.
b)   Hedayat (1977). If n is a prime there exist $n - 3$ mutually orthogonal Knut Vic designs of order n.

Section 3.   Difference sets

In Jungnickel (1980) constructions of difference matrices using planar difference sets and affine difference sets were given. In Jungnickel (1981) similar constructions were given using difference families and partial difference families.   From these we can derive new orthomorphism graphs which will be described in this section. Jungnickel's constructions yield the largest theoretically determined lower bounds for clique numbers of orthomorphism graphs.   Other large lower bounds have been obtained for small groups, using computers. There is only one exception to this rule.   While $\omega(Z_2 \times Z_6)$ was shown to be at least 4 by Bose, Chakravarti, and Knuth (1960) and Johnson, Dulmage, and Mendelsohn (1961), using computers , this has since been shown theoretically by Mills (1977).

A $(v, k, 1)$ - *difference set* in a group G, $|G| = v$, is a k - element subset D of G for which each nonidentity element of G is uniquely expressible in the form $d^{-1}d'$, $d, d' \in D$. A $(v, k, 1)$ - difference set is also called a planar difference set. Let $\{a_1, \dots, a_k\}$ be a $(v, k, 1)$ - difference set in a group G and let $M = \{m_{ij}\}$ be a square matrix of order k with entries $\{1, \dots, k\}$. Define $\theta_M \colon G \to G$ by $\theta_M(a_j^{-1}a_i) = a_j^{-1}a_{m_{ij}}$. We set

$$\mathcal{D}_D(G) = \{\theta_M \colon \theta_M \in \text{Orth}(G)\}.$$ The following theorem characterizes elements of $\mathcal{D}_D(G)$ and adjacency in $\mathcal{D}_D(G)$.

Theorem 5.9. $\theta_M \in \mathcal{D}_D(G)$ if and only if M is a Latin square for which $m_{ii} = i$ for all i. Further, if $\theta_M, \theta_K \in \mathcal{D}_D(G)$ then $\theta_M \sim \theta_K$ if and only if M is orthogonal to K.
Proof. $\theta_M(e) = \theta_M(a_i^{-1}a_i) = a_i^{-1}a_{m_{ii}}$. Thus θ_M will be well defined and will fix e if and only if $m_{ii} = i$ for all i. θ_M will be a bijection if and only if, for all a, the equation $a_j^{-1}a_{m_{ij}} = a$ can be solved uniquely for i and j.   But $a = a_j^{-1}a_r$ for unique values of j and r.   Thus the equation $a_j^{-1}a_{m_{ij}} = a$ uniquely determines j and m_{ij}. It follows that θ_M can be a bijection if and only if each column of M contains each symbol of $\{1, \dots, k\}$ exactly once.   Similarly the mapping $x \to x^{-1}\theta_M(x)$ will be a bijection if and only if, for all a, the equation $a_i^{-1}a_{m_{ij}} = a$ has a unique solution for i and j, if and only if each row of M contains each symbol of $\{1, \dots, k\}$ exactly once.   Hence $\theta_M \in \mathcal{D}_D(G)$ if and only if M is a Latin square for which $m_{ii} = i$ for all i.

If $\theta_M, \theta_K \in \mathcal{D}_D(G)$ then $\theta_M \sim \theta_K$ if and only if, for all a, the equation

$a_{k_{ij}}^{-1} a_{m_{ij}} = a$ can be uniquely solved for i and j, if and only if, for all r and s, the equations $a_{m_{ij}} = a_r$ and $a_{k_{ij}} = a_s$ can be uniquely solved for i and j, if and only if M and K are orthogonal. ∎

Theorem 6 of Jungnickel (1980) is an immediate corollary of Theorem 5.9.

Corollary 5.7. Let q be a prime power and let $v = q^2 + q + 1$. Then
$\omega(\mathrm{Orth}(Z_v)) \geq N(q+1) - 1$. If $q + 1$ is also a prime power then $\omega(\mathrm{Orth}(Z_v)) \geq q - 1$.
Proof. This follows from the well-known existence of a $(v, q + 1, 1)$ - difference set in Z_v, and the fact that if $q + 1$ is also a prime power then $N(q + 1) = q$. ∎

The following table lists some lower bounds for $N(v)$ and $\omega(\mathrm{Orth}(Z_v))$ derived from Corollary 5.7. These provide some of the best lower bounds for clique numbers of orthomorphism graphs of cyclic groups of non-prime order. Some of the lower bounds for $N(v)$ listed in this table were known prior to Jungnickel's work, specifically bounds for $v > 993$. The values of $N(q + 1)$ used in this table can be found in Jungnickel (1990).

q	$q + 1$	v	$N(v) \geq$	$\omega(\mathfrak{O}_D(Z_v)) \geq$
4	5	21	4	3
7	8	57	7	6
16	17	273	16	15
19	20	381	4	3
31	32	993	31	30
49	50	2451	6	5
61	62	3783	4	3
64	65	4161	7	6
67	68	4557	5	4

If H is a subgroup of G then a $(v, k, 1; H)$ - difference set in G, $|G| = v$, is a k - element subset D of G for which each element of $G - H$ is uniquely expressible in the form $d'^{-1}d$, $d, d' \in D$, and no element of H is expressible in this form. A $(v, k, 1; H)$ - difference set is also called an affine difference set. Let $\{a_1, \dots, a_k\}$ be a $(v, k, 1; H)$ - difference set in a group G, let $M = \{m_{ij}\}$ be a square matrix of order k with entries $\{1, \dots, k\}$, and let ϕ be a mapping $H \to H$. Define $\theta_{M, \phi}: G \to G$ by $\theta_{M, \phi}(a_j^{-1}a_i) = a_j^{-1}a_{m_{ij}}$, and $\theta_{M, \phi}(h) = \phi(h)$ for $h \in H$. Set $\mathfrak{I}_{D, H}(G) = \{\theta_{M, \phi}: \theta_{M, \phi} \in \mathrm{Orth}(G)\}$.

Theorem 5.10. $\theta_{M, \phi} \in \mathfrak{I}_{D, H}(G)$ if and only if M is a Latin square with $m_{ii} = i$ for all i and $\phi \in \mathrm{Orth}(H)$. Further, if $\theta_{M, \phi}, \theta_{K, \psi} \in \mathfrak{I}_{D, H}(G)$ then $\theta_{M, \phi} \sim \theta_{K, \psi}$ if and only if M is orthogonal to K and $\phi \sim \psi$.

Proof. $\theta_{M,\phi}(e) = \theta_{M,\phi}(a_i^{-1}a_i) = a_i^{-1}a_{m_{ii}}$. Thus $\theta_{M,\phi}$ will be well defined and will fix e if and only if $m_{ii} = i$ for all i, and ϕ fixes e. $\theta_{M,\phi}$ will be a bijection if and only if ϕ is a bijection and, for all $a \notin H$, the equation $a_j^{-1}a_{m_{ij}} = a$ can be solved uniquely for i and j. But $a = a_j^{-1}a_r$ for unique values of j and r. Thus the equation $a_j^{-1}a_{m_{ij}} = a$ uniquely determines j and m_{ij}. It follows that $\theta_{M,\phi}$ can be a bijection if and only if ϕ is a bijection and each column of M contains each symbol of $\{1, \dots, k\}$ exactly once. Similarly the mapping $x \to x^{-1}\theta_{M,\phi}(x)$ will be a bijection if and only if the mapping $h \to h^{-1}\phi(h)$ is a bijection $H \to H$ and, for all $a \notin H$, the equation $a_i^{-1}a_{m_{ij}} = a$ has a unique solution for i and j. The equation $a_i^{-1}a_{m_{ij}} = a$ has a unique solution for i and j if and only if each row of M contains each symbol of $\{1, \dots, k\}$ exactly once. Hence $\theta_{M,\phi} \in \mathcal{D}_D(G)$ if and only if $\phi \in \mathrm{Orth}(H)$ and M is a Latin square for which $m_{ii} = i$ for all i.

If $\theta_{M,\phi}, \theta_{K,\psi} \in \mathcal{D}_D(G)$ then $\theta_{M,\phi} \sim \theta_{K,\psi}$ if and only if, for all a, the equation $a_{k_{ij}}^{-1}a_{m_{ij}} = a$ can be uniquely solved for i and j and $\phi \sim \psi$, if and only if, for all r and s, the equations $a_{m_{ij}} = a_r$ and $a_{k_{ij}} = a_s$ can be uniquely solved for i and j and $\phi \sim \psi$, if and only if M and K are orthogonal and $\phi \sim \psi$. ∎

An immediate corollary.

Corollary 5.8. Jungnickel (1980, Theorem 7). If q is a prime power and $v = q^2 - 1$ then $\omega(\mathrm{Orth}(Z_v)) \geq \omega(\mathrm{Orth}(Z_{q-1}))$.
Proof. It is well-known that the group Z_v has a $(v, q, 1; Z_{q-1})$ - difference set. ∎

The result of Corollary 5.8 can only be of interest if q is even. In the following table we list some of the consequences of this result.

q	$q-1$	v	$N(v) \geq$	$\omega(\mathcal{I}_{D,H}(Z_v)) \geq$
32	31	3.11.31	30	29
128	127	3.43.127	126	125
512	511	$3^3.7.19.73$	6	5
2048	2047	3.23.89.683	22	21

Let $D = \{D_1, \dots, D_s\}$ be a family of subsets of the element set of a group G, $|G| = v$, and let $K = \{|D_i|: i = 1, \dots, s\}$, $1, 2 \notin K$. We call D a $(v, K, 1)$ - difference family in G if each nonidentity element of G is uniquely expressible in the form $d^{-1}d'$, $d, d' \in D_i$, some i. If $K = \{k\}$ then we may use the term $(v, k, 1)$ - difference family in place of the term $(v, K, 1)$ - difference family. Let $D = \{D_1, \dots, D_s\}$, $D_i = \{a_{ij}: j = 1, \dots, |D_i|\}$, be a $(v, K, 1)$ - difference family in a group G and let $M =$

$\{M_1, \dots, M_s\}$ be a family of square matrices, M_i of order $|D_i|$ with entries $\{1, \dots, |D_i|\}$, the $j$rth element of M_i being denoted m_{ijr}. Define $\theta_M \colon G \to G$ by $\theta_M(a_{ij}^{-1}a_{ir}) = a_{ij}^{-1}a_{im_{ijr}}$. We set $\mathcal{D}_D(G) = \{\theta_M \colon \theta_M \in \mathrm{Orth}(G)\}$. The following theorem is a generalization of Theorem 5.9.

Theorem 5.11. $\theta_M \in \mathcal{D}_D(G)$ if and only if for all i the matrix M_i is a Latin square for which $m_{ijj} = j$ for all j. Further, if $\theta_M, \theta_K \in \mathcal{D}_D(G)$ then $\theta_M \sim \theta_K$ if and only if M_i is orthogonal to K_i for all i.

Proof. This is a fairly straightforward modification of the proof of Theorem 5.9. ∎

Corollary 5.9. Jungnickel (1981). Let q be a prime power and let $v = q^n + \dots + q + 1$, n even. Then $\omega(\mathrm{Orth}(Z_v)) \geq N(q + 1) - 1$ and if $q + 1$ is a prime power then $\omega(\mathrm{Orth}(Z_v)) \geq q - 1$.

Proof. Jungnickel (1981) constructed a $(v, q + 1, 1)$ - difference family in Z_v. The result then follows from Theorem 5.11 and the fact that if $q + 1$ is a prime power then $N(q + 1) = q$. ∎

Let $D = \{D_1, \dots, D_s\}$ be a family of subsets of the element set of a group G, $|G| = v$, and let $K = \{|D_i| \colon i = 1, \dots, s\}$, $1, 2 \notin K$. If H is a subgroup of G then we call D a $(v, K, 1; H)$ - difference family in G if each nonidentity element of $G - H$ is uniquely expressible in the form $d^{-1}d'$, $d, d' \in D_i$, some i, and no element of H can be so expressed. Let $D = \{D_1, \dots, D_s\}$, $D_i = \{a_{ij} \colon j = 1, \dots, |D_i|\}$, be a $(v, K, 1; H)$ - difference family in a group G and let $M = \{M_1, \dots, M_s\}$ be a family of square matrices M_i of order $|D_i|$ with entries $\{1, \dots, |D_i|\}$, the $j$rth element of M_i being denoted m_{ijr}, and let ϕ be a mapping $H \to H$. Define $\theta_{M,\phi} \colon G \to G$ by $\theta_{M,\phi}(a_{ij}^{-1}a_{ir}) = a_{ij}^{-1}a_{im_{ijr}}$ and $\theta_{M,\phi}(h) = \phi(h)$ for $h \in H$. We set $\mathcal{J}_{D,H}(G) = \{\theta_{M,\phi} \colon \theta_{M,\phi} \in \mathrm{Orth}(G)\}$. The following theorem is a generalization of Theorem 5.10.

Theorem 5.12. $\theta_{M,\phi} \in \mathcal{J}_{D,H}(G)$ if and only if for all i the matrix M_i is a Latin square for which $m_{ijj} = j$ for all j and $\phi \in \mathrm{Orth}(H)$. Further, if $\theta_{M,\phi}, \theta_{K,\psi} \in \mathcal{J}_{D,H}(G)$ then $\theta_{M,\phi} \sim \theta_{K,\psi}$ if and only if M_i is orthogonal to K_i for all i and $\phi \sim \psi$.

Proof. This is a fairly straightforward modification of the proof of Theorem 5.10. ∎

Corollary 5.10. Jungnickel (1981). Let q be a prime power and let $v = q^n + \dots + q + 1$, n odd. Then $\omega(\mathrm{Orth}(Z_v)) \geq \omega(\mathrm{Orth}(Z_{q+1}))$ and if $q + 1$ is also a prime power then $\omega(\mathrm{Orth}(Z_v)) \geq q - 1$.

Proof. Jungnickel (1981) constructed a $(v, q + 1, 1)$ - difference family in Z_v. The result then follows from Theorem 5.12 and the fact that if $q + 1$ is also a prime power then $\omega(\mathrm{Orth}(Z_{q+1})) = q - 1$. ∎

Corollary 5.11. Jungnickel (1981). Let q be a prime power and let $v = q^n - 1$. Then $\omega(\mathrm{Orth}(Z_v)) \geq \omega(\mathrm{Orth}(Z_{q-1}))$ and if $q - 1$ is also a prime power then $\omega(\mathrm{Orth}(Z_v)) \geq q - 3$.

Proof. This follows from the existence of a $(v, q, 1)$ - difference family in Z_v, and the fact that if $q - 1$ is also a prime power then $\omega(\mathrm{Orth}(Z_{q-1})) = q - 3$. ∎

Section 4. Maximal sets of orthomorphisms

We know from Theorem 1.8 that the existence of a maximal set of r orthomorphisms of a group G of order n implies the existence of a maximal set of $r + 1$ mutually orthogonal Latin squares of order n. Our first result is a generalization of this theorem to difference matrices. If D is a difference matrix and m a positive integer then mD is the difference matrix $(D \dots D)$, consisting of m consecutive copies of D.

Theorem 5.13. Evans (1991). Let D be an $(n, r; 1, G)$ - difference matrix for which mD is maximal. If there exist $r - 1$ mutually orthogonal Latin squares of order m then there exists a maximal set of $r - 1$ mutually orthogonal Latin squares of order nm.

Proof. If D' is obtained from D by permuting rows and columns, or by multiplying all the elements of any given row of D on the right by a constant or all the elements of any given column of D on the left by a constant, then D' will still be a difference matrix and mD' will still be maximal. Thus we are free to assume that $D = (d_{ij})$, $d_{1j} = e$ for all j. Let $L_k = (a_{ijk})$, $k = 1, \dots, r - 1$, be a set of mutually orthogonal Latin squares of order m, on the symbol set $\{0, \dots, m - 1\}$. Define $\mathfrak{L}_k = (b_{ijk})$ by $b_{ijk} = d_{2i}d_{k+1\,j}$. Then $\mathfrak{L}_1, \dots, \mathfrak{L}_{r-1}$ is a set of mutually orthogonal Latin squares of order n.

Form the Kronecker products $L_1 \times \mathfrak{L}_1, \dots, L_{r-1} \times \mathfrak{L}_{r-1}$. This is a set of mutually orthogonal Latin squares of order mn with entries from the symbol set $\{(i, g): g \in G, i = 0, \dots, m - 1\}$. Suppose that there exists a Latin square M orthogonal to each of $L_1 \times \mathfrak{L}_1, \dots, L_{r-1} \times \mathfrak{L}_{r-1}$. The symbol $(0, e)$ will occur exactly once as an entry in each row and column of M. Let the corresponding cells of $L_1 \times \mathfrak{L}_1$ contain $(a_{i_s j_s 1}, b_{u_s v_s 1})$, $s = 1, \dots, mn$. Define an $(r + 1) \times mn$ matrix $D' = (d_{ij}')$, $d_{1s}' = e$ for $s = 1, \dots, mn$, $d_{k+1\,s}' = d_{k+1\,v_s}$ for $s = 1, \dots, mn$ and $k = 1, \dots, r - 1$, and $d_{r+1\,s}' = d_{2\,u_s}^{-1}$ for $s = 1, \dots, mn$. The sequence $\{(d_{1\,s}')^{-1}d_{r+1\,s}': s = 1, \dots, mn\} = \{d_{2\,u_s}^{-1}: s = 1, \dots, mn\}$ contains each element of G exactly m times as the sequence $\{u_s: s = 1, \dots, mn\}$ contains m copies of each integer $i = 1, \dots, n$. The sequence $\{(d_{k+1\,s})^{-1}d_{r+1\,s}': s = 1, \dots, mn\} = \{(d_{2\,u_s}d_{k+1\,v_s})^{-1}: s = 1, \dots, mn\} = \{b_{u_s v_s k}^{-1}: s = 1, \dots, mn\}$ contains each element of G exactly m times as each element of the symbol set occurs exactly once in the sequence $\{(a_{i_s j_s k}, b_{u_s v_s k}): s = 1, \dots, mn\}$. Thus D' is an

$(n, r + 1; m, G)$ - difference matrix. This contradicts the maximality of mD. ∎

Thus we are led to ask for which $(n, r; 1, G)$ - difference matrices D and which positive integers m is mD a maximal difference matrix? Obvious necessary conditions are that D be maximal and that n not divide m. These are not sufficient however, as if $G = Z_n$, n even, and D is any $(n, 2; 1, G)$ - difference matrix then D will be maximal, but mD will be maximal if and only if m is odd. We will give constructions of maximal difference matrices of the form mD.

Drake (1977) proved that if n does not divide m then the existence of a projective plane of order n and the existence of $n - 1$ mutually orthogonal Latin squares of order m implies the existence of a maximal set of $n - 1$ mutually orthogonal Latin squares of order mn. A straightforward modification of Drake's proof yields the following result.

Theorem 5.14. If D is an $(n, n; 1, G)$ - difference matrix then mD is maximal if and only if n does not divide m.

Theorem 5.14 cannot yield any new parameter pairs for maximal sets of mutually orthogonal Latin squares, but it does give rise to difference matrix proofs of some known results. Corollaries 5.12 and 5.13 follow immediately from Theorems 1.8 and 5.14.

Corollary 5.12. Drake (1977). If q is a prime power, q does not divide m, and there exist $q - 1$ mutually orthogonal Latin squares of order m then there exists a maximal set of $q - 1$ mutually orthogonal Latin squares of order qm.

Corollary 5.13. Bruck (1963). If q is the smallest prime power that divides n then there exists a maximal set of $q - 1$ mutually orthogonal Latin squares of order n.

We next determine some new parameter pairs for maximal sets of mutually orthogonal Latin squares. We will construct, for p a prime, maximal sets of $p - 1$ mutually orthogonal Latin squares of order p^r, and maximal sets of $p - 1$ mutually orthogonal Latin squares of order mp^r, for certain values of m. First we need two lemmas.

Lemma 5.3. Let $n = p^r$, p a prime, and let $\phi: \{0, \dots, p - 1\} \to \{0, \dots, n - 1\}$ be any function satisfying $\phi(i) \equiv i$ modulo p, and let $d_1, \dots, d_{p-1}$ be integers. Then the system of equations $\sum_{i=1}^{p-1} x_i \, \phi(i)^j \equiv d_j$ modulo n, $j = 1, \dots, p - 1$, has a unique solution modulo n.

Proof. For $r = 1$ this is known to be true. Suppose this to be true for $r = k$ and consider the case $r = k + 1$. Set $x_i = y_i + z_i p^k$ where $0 \le y_i < p^k$ and $0 \le z_i < p$.
$\sum_{i=1}^{p-1} x_i \, \phi(i)^j \equiv \sum_{i=1}^{p-1} y_i \, \phi(i)^j + (\sum_{i=1}^{p-1} z_i \, \phi(i)^j) p^k \equiv d_j$ modulo n. Modulo p^k the y_is are uniquely determined. $\sum_{i=1}^{p-1} y_i \, \phi(i)^j \equiv c_j$ modulo n and $c_j \equiv d_j$ modulo p^k.

Thus $(\sum\limits_{i=1}^{p-1} z_i\, \phi(i)^j) p^k \equiv d_j - c_j$ modulo n or $\sum\limits_{i=1}^{p-1} z_i\, \phi(i)^j \equiv (d_j - c_j)/p^k$ modulo p. Thus the z_is are also uniquely determined. Hence the result. ∎

Lemma 5.4. If n is a power of a prime p then $\sum\limits_{i=0}^{n-1} i^{p-1} \equiv (n/p)(p-1)$ modulo n.

Proof. We know this formula to be correct if n is a prime or $p = 2$, so let us assume that $n = p^r$, p an odd prime, $r > 1$, and that we have proved the formula correct for all smaller powers of p. Let g be primitive modulo n. Then

$$\sum_{i=0}^{n-1} i^{p-1} \equiv \sum_{j=1}^{n-(n/p)} (g^j)^{p-1} + \sum_{j=0}^{(n/p)-1} (jp)^{p-1}$$

$$\equiv \sum_{j=1}^{n-(n/p)} g^{j(p-1)} + p^{p-1} \sum_{j=0}^{(n/p)-1} j^{p-1} \text{ modulo } n.$$

If $r = 2$ then $p^{p-1} \equiv 0$ modulo n. If $r > 2$ then, by the inductive hypothesis, $\sum\limits_{j=0}^{(n/p)-1} j^{p-1} \equiv 0$ modulo n/p^2 and so $p^{p-1} \sum\limits_{j=0}^{(n/p)-1} j^{p-1} \equiv 0$ modulo n.

Thus $\sum\limits_{i=0}^{n-1} i^{p-1} \equiv \sum\limits_{j=1}^{n-(n/p)} g^{j(p-1)} \equiv (p-1) \sum\limits_{k=0}^{(n/p)-1} (kp+1)$

$\equiv (p-1) \sum\limits_{k=0}^{(n/p)-1} kp + (p-1)(n/p) \equiv (p-1)(n/p)$ modulo n. ∎

Theorem 5.15. Evans (1991). Let G be the group Z_n, $n = p^r$, p a prime, let $\phi: \{0, \dots, p-1\} \rightarrow \{0, \dots, n-1\}$ be any function satisfying $\phi(i) \equiv i$ modulo p, and suppose that p does not divide m. Let D be the $(n, p; 1, G)$ - difference matrix with ijth. entry $\phi(i-1)j$ modulo n, $i = 1, \dots, p$, $j = 0, \dots, n-1$. Then mD is a maximal difference matrix.

Proof. Let d_{ij} denote the ijth. entry of mD, $i = 1, \dots, p$ and $j = 1, \dots, mn$, and suppose that mD is not maximal. Then we may add an extra row whose entry in the jth. column is d_j, $j = 1, \dots, mn$.

By Lemma 5.3, we may choose $a_1, \dots, a_{p-1}$ to satisfy

$$\sum_{i=1}^{p-1} a_i\, \phi(i)^j \equiv \begin{cases} 0 \text{ modulo } n \text{ for } j < p-1 \\ 1 \text{ modulo } n \text{ for } j = p-1 \end{cases}$$

Then $m(a_1 + \dots + a_{p-1}) \sum\limits_{x=0}^{n-1} x^{p-1} \equiv \sum\limits_{i=2}^{p} a_{i-1} \sum\limits_{j=1}^{mn} (d_j - d_{ij})^{p-1}$

$$\equiv \sum_{i=2}^{p} a_{i-1} \sum_{j=1}^{mn} (d_j - \phi(i-1)d_{2j} / \phi(1))^{p-1}$$

$$\equiv \sum_{j=1}^{mn} \sum_{k=0}^{p-1} \binom{p-1}{k}(-1)^{p-1-k} d_j^k (d_{2j}/\phi(1))^{p-1-k} \sum_{i=2}^{p} a_{i-1} \phi(i-1)^{p-1-k}$$

$$\equiv \sum_{j=1}^{mn} d_j^{p-1} \sum_{i=2}^{p} a_{i-1} + (-1)^{p-1} \sum_{j=1}^{mn} (d_{2j}/\phi(1))^{p-1}$$

$$\equiv m\,(a_1 + \dots + a_{p-1} + \varepsilon) \sum_{x=0}^{n-1} x^{p-1} \text{ modulo } n, \text{ where } \varepsilon = 1 \text{ if } p > 2, -1 \text{ if } p = 2.$$

Therefore $m \sum_{x=0}^{n-1} x^{p-1} \equiv 0$ modulo n, contradicting Lemma 5.4. Hence the result. ∎

In the following corollaries we list several applications of Theorem 5.15.

Corollary 5.14. Let G be a group of order $n = p^r m$, p the smallest prime divisor of n, $r \geq 1$, and p and m relatively prime. Then any $(p-2)$ - clique of $\mathcal{P}(G)$ is a maximal set of orthomorphisms of G.
Proof. This is an immediate consequence of Theorems 5.13 and 5.15. ∎

We should note two special cases of this corollary. From the special case $p = 2$ we can easily derive Hall and Paige's theorem (Theorem 1.20) on the nonexistence of orthomorphisms, and $p = 3$ yields a special case of Theorem 5.4.

Corollary 5.15. Let $n = p^r$, p a prime and $r \geq 1$. Then there exists a maximal set of $p - 1$ mutually orthogonal Latin squares of order n.
Proof. By Theorem 1.8, if there exists a maximal $(n, r; 1, G)$ - difference matrix then there exists a maximal set of $r - 1$ mutually orthogonal Latin squares of order n. ∎

Corollary 5.16. Let $n = p^r m$, p a prime, $r \geq 1$, and p and m relatively prime. If there exist $p - 1$ mutually orthogonal Latin squares of order m then there exists a maximal set of $p - 1$ mutually orthogonal Latin squares of order n.
Proof. This is an immediate consequence of Theorems 5.13 and 5.15. ∎

We list some special cases of Corollaries 5.15 and 5.16.

Example 5.5. If $n = m2^s$, m odd, then there exists a Latin square of order n that has no orthogonal mate. This was first proved by Euler (1779) and has since been rediscovered several times. See Corollary 1.12.

Example 5.6. If $n = m3^s$, m not divisible by 3, then there exists a maximal set of 2 mutually orthogonal Latin squares of order n whenever $m \neq 2, 6$. The special case m = 1, s = 2, can be found in Jungnickel and Grams (1986), where it is proved using a computer search. The case m odd is a special case of Theorem 5.4.

Example 5.7. If $n = m5^s$, m not divisible by 5, then there exists a maximal set of 4 mutually orthogonal Latin squares of order n whenever $m > 52$. It is known that $N(m) \geq 4$ for all $m > 52$. See Dénes and Keedwell (1974, 1991) or Beth, Jungnickel, and Lenz (1985).

Example 5.8. More generally, if $n = mp^s$, m not divisible by p, p a prime, then there exists a maximal set of $p - 1$ mutually orthogonal latin squares of order n whenever m is sufficiently large.

Theorem 5.16. Evans (To appear - b). Let $p \geq 7$ be a prime.

a)   If $p \equiv 3$ modulo 4 then $\mathfrak{C}_2(p)$ contains a maximal set of $(p - 5)/2$ orthomorphisms of $GF(p)^+$.

b)   If $p \equiv 1$ modulo 4 then $\mathfrak{C}_2(p)$ contains a maximal set of $(p - 3)/2$ orthomorphisms of $GF(p)^+$.

Proof. Let $m = (p - 7)/2$ and let $[A, B] \in \mathfrak{C}_2(p) - \mathfrak{C}_1(p)$. By Corollary 3.4, $[A, B]$ is adjacent to precisely m elements of $\mathfrak{C}_1(p)$. Label these elements $[C_1], \dots, [C_m]$. Now $[A, B]$ and $[B, A]$ are both adjacent to $[C_1], \dots, [C_m]$ and $[A, B]$ is adjacent to $[B, A]$ if and only if $p \equiv 1$ modulo 4. Thus $[A, B], [C_1], \dots, [C_m]$ form a $(p - 5)/2$ - clique of $\mathfrak{C}_2(p)$ and if $p \equiv 1$ modulo 4 then we can extend this set using $[B, A]$ to a $(p - 3)/2$ - clique of $\mathfrak{C}_2(p)$. We will prove these cliques of orthomorphisms to be maximal in $\text{Orth}(GF(p)^+)$.

The orthomorphism $[C_i]$ is represented by the orthomorphism polynomial $C_i x$, the orthomorphism $[A, B]$ by the orthomorphism polynomial $\dfrac{(A - B)}{2} x^{(p + 1)/2} + \dfrac{(A + B)}{2} x$, and the orthomorphism $[B, A]$ by the orthomorphism polynomial $\dfrac{(B - A)}{2} x^{(p + 1)/2} + \dfrac{(B + A)}{2} x$.

First we will prove that the only orthomorphisms that can be adjacent to $[C_1], \dots, [C_m]$ are linear orthomorphisms, quadratic orthomorphisms, and their translates. Let $f(x) = a_{1,1} x + a_{2,1} x^2 + \dots + a_{p-1,1} x^{p-1}$ be an orthomorphism polynomial adjacent to $C_1 x, \dots, C_m x$. Thus $f(x) - yx$ is a permutation polynomial of $GF(p)$ for $y = 0, 1, C_1, \dots, C_m$.

Let $f(x)^t \equiv a_{1,t} x + a_{2,t} x^2 + \dots + a_{p-1,t} x^{p-1}$ modulo $x^p - x$. Then, for $t < p - 1$, the coefficient of x^{p-1} in the reduction of $(f(x) - yx)^t$ modulo $x^p - x$ is

$$\sum_{i=0}^{t-1} (-1)^i \binom{t}{i} a_{p-1-i, t-i} y^i.$$

By Theorem 3.12 this coefficient must be congruent to 0 modulo $y(y - 1)(y - C_1) \dots (y - C_m)$ and so $a_{p-1-i, t-i} = 0$ for $0 \leq i \leq t - 1 < m + 2 = (p - 3)/2$. This can be rewritten as follows by setting $k = t - i$.

(*) $a_{(p-1)-i, k} = 0$ for $i \geq 0$, $k \geq 1$, and $k + i \leq (p - 3)/2$.

By setting $k = 1$ in (*) we find that $a_{p-1,1} = a_{p-2,1} = \dots = a_{(p+3)/2,1} = 0$. There are now two cases to consider.

<u>Case 1</u> $a_{(p+1)/2,\ 1} \neq 0$. We may assume that $a_{(p-1)/2,\ 1} = 0$, as this will certainly be true for some translate of $f(x)$. Suppose that $a_{n,\ 1} \neq 0$ for some $n < (p-1)/2$ and that $a_{n',\ 1} = 0$ whenever $n < n' \leq (p-1)/2$.

Considering $f(x)^2$, one sees that $a_{n,\ 1} = a_{n+(p+1)/2,\ 2} / 2a_{(p+1)/2,\ 1}$. But this must be zero if $n + (p+1)/2 \geq p - 1 + 2 - (p-3)/2$, i. e. if $n \geq 2$. Thus $n = 1$ and $f(x)$ represents a quadratic orthomorphism or a translate of a quadratic orthomorphism.

<u>Case 2</u> $a_{(p+1)/2,\ 1} = 0$. Suppose that $a_{n,\ 1} \neq 0$ for some $n < (p+1)/2$ and that $a_{n',\ 1} = 0$ whenever $n < n' \leq (p+1)/2$. Assume $n \neq 1$. Now n cannot divide $p - 1$ and so $p - 1 = qn + r$, $1 \leq r < n$. Thus $a_{qn,\ q} = (a_{n\ 1})^q \neq 0$. Therefore, by (*),
$(p-1) - qn + q \geq (p-1)/2$ and so $qn + r - qn + q \geq (qn + r)/2$ or
$n \leq 2 + r/q < 2 + (n-1)/2$, as $q \geq 2$ and $r \leq n - 1$ and we cannot have both $q = 2$ and
$r = n - 1$ (this would imply that $p = 3n$), and so $n < 3$, a contradiction. Thus $n = 1$ and
$f(x)$ represents a linear orthomorphism.

We have thus shown that any orthomorphism adjacent to $[C_1], \ldots, [C_m]$ must be either a linear orthomorphism, a quadratic orthomorphism, or a translate of a quadratic orthomorphism. Further, Theorem 3.18 implies that we may, without loss of generality, restrict ourselves to linear and quadratic orthomorphisms only.

To complete the proof we need to show that only two nonlinear quadratic orthomorphisms can be adjacent to $[C_1], \ldots, [C_m]$. Suppose that $f(x) = ax^{(p+1)/2} + bx$ and $f'(x) = a'x^{(p+1)/2} + b'x$ are both nonlinear quadratic orthomorphisms adjacent to $[C_1], \ldots, [C_m]$. The coefficients of x^{p-1} in the reductions of $(f(x) - yx)^{(p-1)/2}$ and $(f'(x) - yx)^{(p-1)/2}$ modulo $x^p - x$ are both polynomials in y of degree $m + 2$, and so must be constant multiples of each other, as each is congruent to 0 modulo $y(y - 1)(y - C_1) \ldots (y - C_m)$.

The coefficients of x^{p-1} in the reductions of $(f(x) - yx)^{(p-1)/2}$ and $(f'(x) - yx)^{(p-1)/2}$ modulo $x^p - x$ are

$((p-1)/2)(b - y)^{(p-3)/2}a + \binom{(p-1)/2}{3}(b-y)^{(p-7)/2}a^3$ + terms in y^k, $k < (p-7)/2$,

and $((p-1)/2)(b' - y)^{(p-3)/2}a' + \binom{(p-1)/2}{3}(b'-y)^{(p-7)/2}a'^3$ + terms in y^k,

$k < (p-7)/2$ respectively. Comparing coefficients we see that $a' = \alpha a$, $a'b'^2 = \alpha ab$, for some α, which implies that $b' = b$. Further $a'b'^2 = \alpha ab^2$, and so $a'^3 = \alpha a^3$. Thus $\alpha = \pm 1$ and so $a' = \pm a$. Thus the only orthomorphisms adjacent to $[C_1], \ldots, [C_m]$ are the quadratic orthomorphisms $f(x) = ax^{(p+1)/2} + bx$ and $f'(x) = -ax^{(p+1)/2} + bx$, their translates, and linear orthomorphisms. Hence the result. ∎

An immediate corollary.

Corollary 5.17. Let $p \geq 7$ be a prime.

a)   If $p \equiv 3$ modulo 4 then there exists a maximal set of $(p-3)/2$ mutually orthogonal Latin squares of order p.

b)   If $p \equiv 1$ modulo 4 then there exists a maximal set of $(p - 1)/2$ mutually orthogonal Latin squares of order p.

As examples, Corollary 5.17 establishes the existence of maximal sets of r mutually orthogonal Latin squares of order n for $(n, r) = (7, 2)$, $(11, 4)$, $(13, 6)$, $(17, 8)$, $(19, 8)$, $(23, 10)$, and $(29, 14)$. Of these only the first two parameter pairs are already known, the second pair having been found through a computer search (See Evans and McFarland (1984)).

Section 5.   Complete sets of orthomorphisms

Which groups admit complete sets of orthomorphisms? In this section our main concern will be attempting to answer this question. We will not discuss classes of projective planes that can be constructed from orthomorphisms - those known are translation planes, dual translation planes, and certain derived dual translation planes - except for the special case of planes of prime order. This latter case is of interest as orthomorphisms have been used in the search for non-Desarguesian planes of prime order. These planes have long been conjectured not to exist.

We will prove some non-existence results derived from proofs of the non-existence of certain Generalized Hadamard matrices. In particular we will present de Launey's (1984a and 1984b) and Woodcock's (1986) results on the non-existence of certain generalized Hadamard matrices. This combined with Theorem 1.10 yields non-existence results for complete sets of orthomorphisms of certain groups.

There is only one class of groups that are known to admit complete sets of orthomorphisms, the class of elementary abelian groups.

Theorem 5.17. Elementary abelian groups admit complete sets of orthomorphisms.
Proof. See Example 1.2. ∎

We now present two non-existence results for complete sets of orthomorphisms.

Theorem 5.18. Let G be a group of order n. If $n \equiv 1$ or 2 modulo 4, and n is not the sum of two squares then G cannot admit a complete set of orthomorphisms.
Proof. This is a special case of the Bruck - Ryser Theorem. See Dembowski (1968). ∎

Theorem 5.19. A group with a nontrivial cyclic Sylow 2-subgroup cannot admit a complete set of orthomorphisms.
Proof. This is a consequence of Theorem 1.20. ∎

Studnicka (1972) proved that Cartesian planes of order $2p^n$ did not exist, a special case of Theorem 5.19. The next theorem is mainly the result of computer studies of orthomorphism graphs, that will be summarized in the next chapter.

Theorem 5.20. Except for the elementary abelian groups, no group of order less than 20 can admit a complete set of orthomorphisms.

By Theorem 1.10 any nonexistence results for generalized Hadamard matrices give rise to nonexistence results for complete sets of orthomorphisms. We give two such results here. We shall use the symbol $GH(n, G)$ to denote an $n \times n$ generalized Hadamard matrix over G.

Theorem 5.21. De Launey (1984a).

i) Let $G = Z_3$ and $n = 3^t p_1{}^{k_1} \ldots p_m{}^{k_m}$, where $p_1, \ldots, p_m$ are distinct odd primes. If $p_i \equiv 5$ modulo 6 and k_i is odd, for some i, then no $GH(n, G)$ exists.

ii) Let $G = Z_5$ and $n = 3^t 7^k m$. If m is odd, m is relatively prime to 21, and one of t, k is odd, then no $GH(n, G)$ exists.

iii) Let $G = Z_7$ and $n = 3^t m$. If m is relatively prime to 3, and t and m are both odd, then no $GH(n, G)$ exists.

The following are immediate consequences.

Corollary 5.18. Let G admit a homomorphism onto Z_3, and $|G| = 3^t p_1{}^{k_1} \ldots p_m{}^{k_m}$, where $p_1, \ldots, p_m$ are distinct odd primes. If $p_i \equiv 5$ modulo 6 and k_i is odd, for some i, then G does not admit a complete set of orthomorphisms.

In particular the groups $Z_{15}, Z_{33}, Z_{45}, Z_{99}, Z_{105}$, and any group of these orders, with a normal subgroup of index 3, cannot admit a complete set of orthomorphisms.

Corollary 5.19. Let G admit a homomorphism onto Z_5, and $|G| = 3^t 7^k m$. If m is odd, m is relatively prime to 21, and one of t, k is odd, then G does not admit a complete set of orthomorphisms.

Corollary 5.20. Let G admit a homomorphism onto Z_7, and $|G| = 3^t m$. If m is relatively prime to 3, and t and m are both odd, then G does not admit a complete set of orthomorphisms.

For the special case G an abelian group of order n the non-existence results for a $GH(n, G)$ of Theorem 5.21 can be generalized.

Theorem 5.22. De Launey (1984b). Let G be an abelian group of order n and let $p \neq 2$ be a prime that divides n. If there exists an integer $m \not\equiv 0$ modulo p, m dividing the square free part of n, and the order of m modulo p being even, then no $GH(n, G)$ exists.

Theorem 5.23. Woodcock (1986). If G admits a homomorphism onto Z_3 and $|G| \equiv 15$ modulo 18 then G does not admit a complete set of orthomorphisms.

Woodcock actually proved that Z_n cannot admit a complete set of orthomorphisms if $|G| \equiv 15$ modulo 18, a special case of Theorem 5.22 (see Jungnickel (1990)). Woodcock's proof carries over, without modification, to a proof of Theorem 5.23. As special cases of Theorem 5.23 we see that Z_{15}, Z_{33}, and Z_{51}, the only groups of these orders, do not admit complete sets of orthomorphisms. These particular cases also follow from Corollary 5.18.

We next turn to the study of affine (equivalently projective) planes of prime order. It has long been conjectured that all affine planes of prime order must be Desarguesian. A weaker conjecture is that all Cartesian planes of prime order must be Desarguesian. By Corollary 1.5 the study of Cartesian planes of prime order p reduces to the study of $(p - 2)$ - cliques of $\text{Orth}(Z_p)$. We know that $\mathfrak{C}_1$ forms a $(p - 2)$ - clique of $\text{Orth}(Z_p)$. By Theorem 2.1 the existence of any other $(p - 2)$ - clique of $\text{Orth}(Z_p)$ would imply the existence of a non - Desarguesian plane of order p. Johnson, Dulmage and Mendelsohn (1961) showed that $\text{Orth}(Z_{11})$ contained only one 9 - clique, $\mathfrak{C}_1$. Cates and Killgrove (1981) reported that no non - Desarguesian affine planes admitting translations could exist of orders 11 and 13. Evans and Mcfarland (1984) studied the structure of $\text{Orth}(Z_{11})$, confirming the two previous reports. All of these proofs required the use of computers.

Mendelsohn and Wolk (1985) asked if $\mathfrak{C}_2(p)$ could contain a $(p - 2)$ - clique other than $\mathfrak{C}_1(p)$. They showed, using a computer search, that the only $(p - 2)$ - clique contained in $\mathfrak{C}_2(p)$ for $p = 13$ and 17 is $\mathfrak{C}_1(p)$. Subsequently Evans (1987a) extended their result to all odd primes less than or equal to 47 using a graphical algorithm and hand calculations. We will prove that for all odd primes the only $(p - 2)$ - clique in $\mathfrak{C}_2(p)$ is $\mathfrak{C}_1(p)$, thus solving the problem posed by Mendelsohn and Wolk. The solution follows from a theorem of Carlitz.

Theorem 5.24. Carlitz (1960). Let f be a permutation polynomial of $GF(p^n)$ for which $(f(a) - f(b))/(a - b)$ is a non-zero square for all a, b, $a \neq b$. Then $f(x) = cx^{p^i} + d$, c a non-zero square, and $0 \leq i < n$.

Proof. We shall prove this for the special case $n = 1$. Define $f'(x) = f(x) - f(0)$. Let e be a nonsquare and define a polynomial $h(x)$ by $h(x) = ef'(x)$. Then $h(x)$ is an orthomorphism polynomial and is adjacent to $[A]$ for all squares A. Thus $h(x)$ is adjacent to $(p - 5)/2$ linear orthomorphisms. In the proof of Theorem 5.16 we established that only linear orthomorphisms or quadratic orthomorphisms and their translates can be adjacent to $(p - 7)/2$ or more linear orthomorphisms. Further, by Corollary 3.4, nonlinear quadratic orthomorphisms and their translates are adjacent to precisely $(p - 7)/2$ linear orthomorphisms. Thus $h(x)$ must be linear. Hence $f(x) = cx + d$, and c must obviously be a nonzero square. For the proof of the general case see Carlitz (1960). ∎

We obtain the solution to Mendelsohn and Wolk's problem as a corollary.

Corollary 5.21. Evans (1989a). The only $(p - 2)$-clique in $\mathfrak{C}_2(p)$ is $\mathfrak{C}_1(p)$.
Proof. Let $[A_1, B_1], \ldots, [A_{p-2}, B_{p-2}]$ be a $(p - 2)$ - clique in $\mathfrak{C}_2(p)$. Define $f: GF(p) \to GF(p)$ by $f(0) = 0$, $f(1) = 1$, and $f(A_i) = B_i$ for all i. By Theorem 3.1, f satisfies the conditions of Theorem 5.24 and so $A_i = B_i$ for all i. ∎

More generally.

Corollary 5.22. Let $[A_1, B_1], \ldots, [A_{q-2}, B_{q-2}]$ be a $(q-2)$-clique in $\mathfrak{C}_2(q)$, $q = p^n$. Then $B_i = A_i^{p^j}$, all i, some j, $1 \leq j < n$.
Proof. Similar to the proof of Corollary 5.21. ∎

Corollary 5.21 has the following geometric consequence.

Theorem 5.25. If an affine plane of prime order p admits translations and also a homology, with axis l_∞ and order $(p-1)/2$, then it is Desarguesian.
Proof. Let A be an affine plane of prime order p admitting translations and also a homology η, with axis l_∞ and order $(p-1)/2$. We may coordinatize A using Hall's procedure as outlined in Dembowski (1968, p.127) using (∞) (Dembowski uses the symbol v) as the center of a translation and $(0, 0)$ as the center of η. A is thus coordinatized by a Cartesian group (See Dembowski (1968, p.130, 22(a))), the points of A are represented by ordered pairs of elements of Z_p and the lines of A are represented by the equations $x = \text{constant}$, $y = \text{constant}$, $y = x + b$ and $y = \theta_i(x) + b$ for $i = 1, \ldots, p-2$. The mappings $\theta_1, \ldots, \theta_{p-2}$ form a $(p-2)$-clique of Orth(Z_p) (See Johnson, Dulmage and Mendelsohn (1961)). It is an exercise to show that η maps the point (x, y) to the point (gx, gy), $g \in GF(p)^*$, $|g| = (p-1)/2$. Now η fixes the lines $y = \theta_i(x)$ for $i = 1, \ldots, p-2$ and so $\theta_i(gx) = g\theta_i(x)$ or $\theta_i(g^r h) = \theta_i(h)h^{-1}(g^r h)$. As $\{g^r; r = 1, \ldots, (p-1)/2\}$ is the set of squares of GF(p), $\theta_1, \ldots, \theta_{p-2}$ are in $\mathfrak{C}_2$. Hence, by Corollary 5.21, $\theta_1, \ldots, \theta_{p-2}$ are in $\mathfrak{C}_1$, and A is thus Desarguesian. ∎

Chapter 6: Data for small groups

Section 1. Introduction.

In this chapter we will present what is known about Orth(G), for G of small order. Our knowledge about the structure of these graphs is mostly the result of computer searches. The first attempt to systematically study cliques of orthomorphism graphs of abelian groups of small order was by Johnson, Dulmage, and Mendelsohn (1961). There then followed studies by Chang, Hsiang, and Tai (1964) and Chang and Tai (1964). The latter authors used exhaustive hand computations, rather than computer searches.

We will show that, for elementary abelian groups, much can be gleaned using (σ, ε) - systems (see Chapter 2), permutation polynomials (see Chapter 3), and cyclotomy (see chapter 3).

We know that in principle, all orthomorphisms of an elementary abelian group of order q can be found by solving all inequivalent (σ, ε)-systems of order at most $q - 1$. As the number of inequivalent (σ,ε)-systems increases somewhat dramatically with the order, this method can be used to calculate all orthomorphisms of elementary abelian groups of very small order only.

The following theorem will enable us to use the (σ, ε) - systems that we have already solved to determine all orthomorphisms of elementary abelian groups of order 8 or less.

Theorem 6.1. Suppose θ is an orthomorphism of $GF(q)^+$, $q > 3$. Then there exists a (σ, ε)-system of order at most $q - 3$ that yields θ.

Proof. As $\theta(x_i)/x_i \neq 0$, 1 there exist j, k, and a such that $\theta(x_j) = ax_j$ and $\theta(x_k) = ax_k$. Define σ by $x_{\sigma(i)} = \theta(x_i)/a$ and ε by $x_{\sigma(i)} = (\theta(x_i) - x_i)/(a - 1)$ to complete the proof. ∎

Thus to determine all orthomorphisms of $GF(q)^+$ it is sufficient to solve all (σ, ε)-systems of order at most $q - 3$. The results of Chapter 2 enable us to solve all (σ, ε)-systems of order at most 5 and hence we can calculate all orthomorphisms of $GF(q)^+$ for $q \leq 8$. For $q = 3$, 4, and 5, as there are no systems of orders 1 or 2, we obtain only linear orthomorphisms.

For $q = 7$ the results are more interesting. We obtain 5 linear orthomorphisms. No system of order 4 yields orthomorphisms of Z_7. 14 non-linear orthomorphisms are obtained from the (σ, ε)-system with $\sigma = (1\ 2\ 3)$ and $\varepsilon = (1\ 3\ 2)$. These extra orthomorphisms are of degree 0 in Orth(Z_7). This is easily seen using our knowledge of quadratic orthomorphisms. From Example 3.1 we know that $\mathfrak{C}_2(7) - \mathfrak{C}_1(7) = \{[3, 5], [5, 3]\}$. From the translates of these quadratic orthomorphisms we obtain the 14 non-linear orthomorphisms of Z_7, which cannot be adjacent to linear orthomorphisms by Corollary 3.4, or to each other by Corollary 3.1 iii) and Theorem 3.18 ii).

The above results can also be obtained easily using the results on permutation polynomials given in Theorems 3.14 and 3.15

Section 2. Groups of order 8.

There are 5 groups of order 8. The group Z_8 has no orthomorphisms by Theorem 1.20.

The groups D_4, the dihedral group, and Q_8, the quaternion group, have 48 orthomorphisms each, no two of which are adjacent. This was first reported in Chang, and Tai (1964). Jungnickel and Grams (1986) confirmed that the clique number of each of these orthomorphism graphs is one. Using Cayley, Evans and Perkel recently verified Chang and Tai's results.

The group $Z_2 \times Z_4$ has 48 orthomorphisms. This was first established by Johnson, Dulmage, and Mendelsohn (1961), who partitioned this set of orthomorphisms into 24 adjacent pairs. Chang, Hsiang, and Tai (1964) also found $Z_2 \times Z_4$ to have 48 orthomorphisms and the clique number of it's orthomorphism graph to be 2. Jungnickel and Grams (1986) also found $\omega(Z_2 \times Z_4)$ to be 2. In what follows we present the results of a computer search, done by Evans and Perkel, using Cayley. We may think of the elements of this group as the ordered pairs (i, j), $i \in Z_2$ and $j \in Z_4$. For simplicity we shall write ij in place of (i, j). The following two tables list some of the orthomorphisms. Those not listed are translates of these.

x	00	10	01	11	02	12	03	13
$\alpha_1(x)$	00	13	11	02	03	10	12	01
$\alpha_2(x)$	00	03	12	13	01	02	11	10
$\alpha_3(x)$	00	01	13	12	11	10	02	03
$\alpha_4(x)$	00	11	10	03	13	02	01	12
$\beta_1(x)$	00	12	13	01	03	11	10	02
$\beta_2(x)$	00	02	12	10	11	13	01	03
$\beta_3(x)$	00	12	11	03	13	01	02	10
$\beta_4(x)$	00	02	10	12	01	03	13	11
$\gamma_1(x)$	00	11	10	13	12	03	02	01
$\gamma_2(x)$	00	01	03	12	10	11	13	02
$\gamma_3(x)$	00	13	02	03	12	01	10	11
$\gamma_4(x)$	00	03	11	02	10	13	01	12

x	00	10	01	11	02	12	03	13
$\delta_1(x)$	00	01	10	03	12	13	02	11
$\delta_2(x)$	00	11	03	02	10	01	13	12
$\delta_3(x)$	00	03	02	13	12	11	10	01
$\delta_4(x)$	00	13	11	12	10	03	01	02
$\zeta_1(x)$	00	03	13	02	12	11	01	10
$\zeta_2(x)$	00	13	02	01	10	03	12	11
$\zeta_3(x)$	00	01	03	10	12	13	11	02
$\zeta_4(x)$	00	11	12	13	10	01	02	03
$\eta_1(x)$	00	11	13	10	12	03	01	02
$\eta_2(x)$	00	01	02	13	10	11	12	03
$\eta_3(x)$	00	13	03	02	12	01	11	10
$\eta_4(x)$	00	03	12	01	10	13	02	11

Each orthomorphism of $Z_2 \times Z_4$ has degree 2 and so this orthomorphism graph is the disjoint union of cycles; the disjoint union of twelve 4-cycles in fact. The orthomorphisms α_1, α_2, α_3, and α_4 form a 4-cycle, $\alpha_i \sim \alpha_{i+1}$, the subscripts being added modulo 4. Similarly for β_1, β_2, β_3, and β_4; γ_1, γ_2, γ_3, and γ_4; δ_1, δ_2, δ_3, and δ_4; ζ_1, ζ_2, ζ_3, and ζ_4; and η_1, η_2, η_3, and η_4.

There are three further 4-cycles that are translates of the 4-cycle α_1, α_2, α_3, and α_4, and three further 4-cycles that are translates of the 4-cycle β_1, β_2, β_3, and β_4. Any translation that fixes one of these cycles must fix each vertex of the cycle. Thus each of these 32 orthomorphisms is fixed by a translation of order 2. Each of the remaining four 4-cycles is fixed by all translations. In fact the corresponding orbits under translations are $\{\gamma_1, \gamma_3\}$, $\{\gamma_2, \gamma_4\}$, $\{\delta_1, \delta_3\}$, $\{\delta_2, \delta_4\}$, $\{\zeta_1, \zeta_3\}$, $\{\zeta_2, \zeta_4\}$, $\{\eta_1, \eta_3\}$, and $\{\eta_2, \eta_4\}$.

Now we turn to the group $GF(8)^+$. By Theorem 6.1, to find all orthomorphisms of $GF(q)^+$ by solving (σ, ε) - systems, it is sufficient to solve all (σ, ε) - systems of order at most $q - 3$. Similarly to find all orthomorphism polynomials of $GF(q)^+$, it is sufficient, by Theorem 3.13, to find all orthomorphism polynomials of degree at most $q - 3$. In Chapter 2 we solved all (σ, ε) - systems of order at most 5, and in Chapter 3 we found all orthomorphism polynomials of degree at most 5. Thus we already know all orthomorphisms of $GF(8)^+$, though not of $GF(9)^+$.

Johnson, Dulmage, and Mendelsohn (1961) determined all the 6-cliques of $Orth(GF(8)^+)$, and Chang, Hsiang, and Tai (1964) listed representatives for each orbit of $Orth(GF(8)^+)$ under the action of known congruences. To understand the structure of

Orth(GF(8)$^+$) we will utilize knowledge that we have gleaned through solving (σ, ε) - systems, from construction orthomorphism polynomials, and from our knowledge of the structure of GL(3, 2).

By solving (σ, ε) - systems, we obtain 6 linear orthomorphisms and 42 non-linear orthomorphisms. These are described in Example 2.3. With a, b, and c as in that example it is easy to show that the values of a and b uniquely determine the orthomorphism. In fact, we may let $\{\phi_{a, b}: a \neq 0, 1, b \neq 0\}$ denote the set of orthomorphisms obtained from this example. Then $\phi_{a, b} = ax$ if $x \in \{0, b, ab, ab + b\}$, $a(x + ab + b)$ otherwise.

As $\phi_{a, b}(x) = ax$ if and only if $x \in <b, ab>$, a subgroup of order 4 of GF(8)$^+$, and $\phi_{a, B}(x) = ax$ if and only if $x \in <B, aB>$, another subgroup of order 4 of GF(8)$^+$, there must exist $x \neq 0$ for which $\phi_{a, b}(x) = ax = \phi_{a, B}(x)$. Hence $\phi_{a, b}$ cannot be adjacent to $\phi_{a, B}$.

Further adjacencies and non-adjacencies can be determined by brute force calculations. Determining which linear orthomorphisms are adjacent to a given orthomorphism is a little simpler.

Theorem 6.2. Evans (1987b). Let $\theta(x_i) = ax_{\sigma(i)}$ and suppose $b \neq a$. Then θ is adjacent to the linear orthomorphism $[b]$ if and only if the mapping $x_i \rightarrow (ax_{\sigma(i)} - bx_i)/(a - b)$ is a permutation of the set $\{x_i: i \neq \sigma(i)\}$.
Proof. Routine calculation. ∎

As a consequence of this we obtain.

Corollary 6.1. Evans (1987b). Let $\{a, y_1, \dots, y_{q-1}\}$ and $\{a, z_1, \dots, z_{q-1}\}$ be solutions, yielding orthomorphisms, of the same (σ, ε) - system and suppose that this system has rank two. If θ and ϕ are the corresponding orthomorphisms then $\theta \sim [b]$ if and only if $\phi \sim [b]$.
Proof. We must have $z_i = cy_i + d$ for $i = 1, \dots, n$ for some $c, d \in$ GF(q). Thus $(az_{\sigma(i)} - bz_i)/(a - b) = c(ay_{\sigma(i)} - by_i)/(a - b) + d$ and the result then follows by Theorem 6.2. ∎

For the case $q = 8$, we can show that $\phi_{a, b}$ is adjacent to the linear orthomorphism $[d]$ if and only if $d = a^2$.

All possible orthomorphism polynomials of GF(8) are listed in Theorem 3.15 a) i), and a) ii), and all normalized permutation polynomials of GF(8) are listed in Theorem 3.14 a) i), a) ii), a) vi), a) vii), and a) xii). From this we can, in principle, determine the structure of Orth(GF(8)$^+$). The next theorem describes, in polynomial form, the elements and adjacencies in Orth(GF(8)$^+$).

Theorem 6.3. For the orthomorphism graph of $GF(8)^+$ the following hold.

i) $\mathfrak{C}_1 = \{Ax: A \neq 0, 1\}$.

ii) $\mathfrak{C}_7 - \mathfrak{C}_1 = \{Ax^4 + Bx^2 + Cx: A \neq 0, \text{ and } Ax^3 + Bx + D \text{ is irreducible over } GF(8)$
for $D = C, C - 1\}$.

iii) If $Ax, Bx \in \mathfrak{C}_1$ then $Ax \sim Bx$ if and only if $A \neq B$.

iv) If $Ax \in \mathfrak{C}_1$ and $Bx^4 + Cx^2 + Dx \in \mathfrak{C}_7 - \mathfrak{C}_1$ then $Ax \sim Bx^4 + Cx^2 + Dx$ if and
only if $Bx^3 + Cx + D - A$ is irreducible over $GF(8)$.

v)   If $Ax^4 + Bx^2 + Cx$ and $Dx^4 + Ex^2 + Fx \in \mathfrak{C}_7 - \mathfrak{C}_1$ then
$Ax^4 + Bx^2 + Cx \sim Dx^4 + Ex^2 + Fx$ if and only if $A = D, B = E$ and $C \neq F$, or
$A \neq D$ and $(A - D)x^3 + (B - E)x^2 + C - F$ is irreducible over $GF(8)$.

Proof. Routine computations. ∎

In particular we observe that all 48 orthomorphisms of $GF(8)^+$ are also automorphisms
and so $\text{Orth}(GF(8)^+) = \mathcal{Q}(GF(8)^+)$. Now $\text{Aut}(GF(8)^+) \cong GL(3, 2)$, the simple group of order
168. The elements of $\text{Aut}(GF(8)^+)$ of order 7 must be fixed point free and, as $\text{Aut}(GF(8)^+)$
has 8 Sylow 7-subgroups, the 48 orthomorphisms of $GF(8)^+$ are precisely the 48 elements of
$\text{Aut}(GF(8)^+)$ of order 7. In each of 8 the Sylow 7-subgroups of $\text{Aut}(GF(8)^+)$ the non-identity
elements form a 6-clique of $\text{Orth}(GF(8)^+)$. Let us label these 6-cliques $\mathcal{A}_1, \dots, \mathcal{A}_8$. Now
$GF(8)^+$ can be regarded as the additive group of several different but isomorphic fields, and
each of $\mathcal{A}_1, \dots, \mathcal{A}_8$ is a clique of linear orthomorphisms with respect to one of these fields.
As each non-linear orthomorphism of $GF(8)^+$ is adjacent to precisely one linear orthomorphism
of $GF(8)^+$, it follows that each element of $\mathcal{A}_i$ is adjacent to exactly one element of $\mathcal{A}_j$, for
$i \neq j$. This implies that each vertex of the orthomorphism graph of $GF(8)^+$ has degree 12.
The only non-linear orthomorphisms adjacent to the linear orthomorphism $[a^2]$ are
$\{\{\phi_{a,b}: b \neq 0\}\}$, which are pairwise non-adjacent, and so any 3-clique must be contained in
some $\mathcal{A}_i$. Thus the maximal cliques of $\text{Orth}(GF(8)^+)$ have sizes 2 and 6 only. This last fact
was found by Jungnickel and Grams (1986), as the result of a computer search.

Section 3.   Groups of order 9.

There are two groups of order 9, Z_9 and $Z_3 \times Z_3 \cong GF(9)^+$. Johnson, Dulmage, and
Mendelsohn (1961) found that Z_9 has 225 orthomorphisms. Chang, Hsiang, and Tai (1964)
reported that $\omega(Z_9) = 1$: This was confirmed by Jungnickel and Grams (1986). A complete
list of orthomorphisms of Z_9 can be found in Hsu (1980, Appendix II), where they are
presented as the presentation functions of cyclic neofields of order 10.

Johnson, Dulmage, and Mendelsohn (1961) found all the complete sets of orthomorphisms of GF(9)$^+$: These are all contained in $\mathfrak{C}_4(9)$. More information about the structure of Orth(GF(9)$^+$) can be found in Chang, Hsiang, and Tai (1964). Jungnickel and Grams (1986) found Orth(GF(9)$^+$) to have maximal cliques of sizes 2, 4, and 7 only. We will list, modulo translates, all orthomorphism polynomials of GF(9) of degree less than 6, and also gives partial information about the structure of Orth(GF(9)$^+$). This is presented in the next theorem, which can be found in Evans (1989c).

Theorem 6.4. Let GF(9) = $\{a + ib: a, b \in$ GF(3), $i^2 + 1 = 0\}$. Then the following hold for the orthomorphism graph of GF(9)$^+$.

i) $\mathfrak{C}_1 = \{Ax: A \neq 0, 1\}$.

ii) $\mathfrak{C}_2 - \mathfrak{C}_1 = \{Ax^5: A = \pm i\} \cup \{Ax^5 + Bx: A = \pm i, B = \pm 1\}$.

iii) $\mathfrak{C}_4 - \mathfrak{C}_2 = \{Ax^3: A \text{ a non-square}\} \cup \{Ax^3 + x: A \text{ a non-square}\} \cup$
$\{Ax^3 + Bx: AB \text{ and } A(B - 1) \text{ non-squares}\}$.

iv) The elements of $\mathfrak{C}_8 - \cup \{T_g[\mathfrak{C}_4]: g \in$ GF(9)$\}$ are represented by orthomorphism polynomials of degree 6.

v) If $Ax, Bx \in \mathfrak{C}_1$ then $Ax \sim Bx$ if and only if $A \neq B$.

vi) If $Ax^5 + Bx \in \mathfrak{C}_2 - \mathfrak{C}_1$ and $Cx \in \mathfrak{C}_1$ then $Ax^5 + Bx \sim Cx$ if and only if $C = -1$.

vii) If $Ax^3 + Bx \in \mathfrak{C}_4 - \mathfrak{C}_2$ and $Cx \in \mathfrak{C}_1$ then $Ax^3 + Bx \sim Cx$ if and only if $B = C$ or at least one of $B(B - C)$, $(B - 1)(B - C)$ is a square.

viii) If $Ax^3 + Bx$ and $Cx^3 + Dx$ are two distinct elements of $\mathfrak{C}_4 - \mathfrak{C}_2$ then $Ax^3 + Bx \sim Cx^3 + Dx$ if and only if $(A - C)(B - D)$ is zero or a non-square.

ix) No element of $\mathfrak{C}_2 - \mathfrak{C}_1$ is adjacent to any element of $\mathfrak{C}_4 - \mathfrak{C}_2$.
Proof. This follows from Theorems 3.14, 3.15, and 3.16 by routine computations. ∎

Note that $\mathcal{Q}(GF(9)^+) = \mathfrak{C}_1 \cup (\mathfrak{C}_4 - \mathfrak{C}_2)$ and $|\mathcal{Q}(GF(9)^+)| = 27$. Theorem 6.4 does not list all orthomorphism polynomials of GF(9), even modulo translates, as there are many orthomorphism polynomials of GF(9) of degree 6. Chang, Hsiang, and Tai (1964) found GF(9)$^+$ to have a total of 249 orthomorphisms. This has been confirmed by Evans and Perkel, using Cayley.

Section 4. The cyclic group of order 11.

Next we turn to the group Z_{11}. We will first present computer generated data for this group. Our work so far will enable us to better understand this data. However, our results so far are not sufficiently powerful to permit us to completely explain the results.

Singer (1960) found implicitly that Orth(Z_{11}) has 3441 orthomorphisms. Under the action of the congruence group generated by H_α, T_g, R, and I, these orthomorphisms lie in 11 orbits, 1 orbit of length 3, 1 of length 6, 2 of length 66, 4 of length 330, and 3 of length 660. Singer's list of representatives for these orbits is given in the following table.

x	0	1	2	3	4	5	6	7	8	9	10	|Orbit|
$\theta_1(x)$	0	2	4	6	8	10	1	3	5	7	9	3
$\theta_2(x)$	0	3	6	9	1	4	7	10	2	5	8	6
$\theta_3(x)$	0	2	4	8	3	9	1	10	5	7	6	66
$\theta_4(x)$	0	2	5	8	3	1	10	9	6	4	7	66
$\theta_5(x)$	0	2	4	6	9	1	10	5	3	8	7	330
$\theta_6(x)$	0	2	4	7	9	1	5	10	3	6	8	330
$\theta_7(x)$	0	2	5	8	10	9	3	6	4	7	1	330
$\theta_8(x)$	0	2	5	9	6	10	4	3	1	8	7	330
$\theta_9(x)$	0	2	4	7	1	10	9	6	3	5	8	660
$\theta_{10}(x)$	0	2	4	7	1	10	9	5	3	8	6	660
$\theta_{11}(x)$	0	2	4	7	1	10	9	3	6	8	5	660

3441

Johnson, Dulmage and Mendelsohn (1961) again found $\text{Orth}(Z_{11})$ consisted of 3441 orthomorphisms and contained only one 9-clique. Chang, Hsiang, and Tai (1964) found no edges in $\text{Orth}(Z_{11})$ outside of this 9-clique. The fact that $\text{Orth}(Z_{11})$ has only one 9-clique was rediscovered by Cates and Killgrove (1981).

The structure of this graph was determined by Evans and McFarland (1984), using a computer. They showed that of the 3441 orthomorphisms, 2640 of these have degree 0, 660 have degree 3, 135 have degree 8, and 6 have degree 162. The orthomorphisms of degree 162 are the linear orthomorphisms $\{[A]: A = 3, 4, 5, 7, 8, \text{ and } 9\}$. The remaining linear orthomorphisms have degree 8 and are adjacent only to linear orthomorphisms. The neighborhood of an orthomorphism of degree 3 consists of an independent set of orthomorphisms, 2 of degree 3, and 1 of degree 162. The neighborhood of a non-linear orthomorphism of degree 8 is the cube graph, 6 of its vertices being non-linear orthomorphisms of degree 8, and 2 of its vertices being linear orthomorphisms of degree 162.

How close are we to being able to explain these results? First let us note the actions of the congruences I and R on $\mathfrak{C}_1$. These are described in the following diagrams.

$$
\begin{array}{ccccccc}
 & & R & & & I & \\
3 & & \leftrightarrow & 9 & & \leftrightarrow & 5 \\
\updownarrow\,I & & & & & & \updownarrow\,R \\
4 & & \leftrightarrow & 8 & & \leftrightarrow & 7 \\
 & & R & & & I &
\end{array}
$$

$$
\begin{array}{ccccc}
& R & & I & \\
2 & \leftrightarrow & 10 & \leftrightarrow & 10 \\
\updownarrow\ I & & & & \updownarrow\ R \\
6 & \leftrightarrow & 6 & \leftrightarrow & 2 \\
& R & & I &
\end{array}
$$

Thus the orbit of length 3 consists of the linear orthomorphisms {[a]: a = 2, 6, and 10}, and the orbit of length 6 consists of the linear orthomorphisms {[a]: a = 3, 4, 5, 7, 8, and 9}.

The adjacencies in $\mathfrak{C}_2$ - $\mathfrak{C}_1$ are given in the following table.

Orthomorphism	Neighbors
[2, 6]	[3, 9], [9, 3], [4, 5], [5, 4], [10, 2], [6, 10], [7], [8]
[6, 2]	[3, 9], [9, 3], [4, 5], [5, 4], [2, 10], [10, 6], [7], [8]
[2, 10]	[4, 5], [5, 4], [7, 8], [8, 7], [6, 2], [10, 6], [3], [9]
[10, 2]	[4, 5], [5, 4], [7, 8], [8, 7], [2, 6], [6, 10], [3], [9]
[3, 9]	[2, 6], [6, 2], [6, 10], [10, 6], [5, 4], [8, 7], [5], [7]
[9, 3]	[2, 6], [6, 2], [6, 10], [10, 6], [4, 5], [7, 8], [5], [7]
[4, 5]	[2, 6], [6, 2], [2, 10], [10, 2], [9, 3], [7, 8], [8], [9]
[5, 4]	[2, 6], [6, 2], [2, 10], [10, 2], [3, 9], [8, 7], [8], [9]
[6, 10]	[3, 9], [9, 3], [7, 8], [8, 7], [2, 6], [10, 2], [4], [5]
[10, 6]	[3, 9], [9, 3], [7, 8], [8, 7], [6, 2], [2, 10], [4], [5]
[7, 8]	[2, 10], [10, 2], [6, 10], [10, 6], [9, 3], [4, 5], [3], [4]
[8, 7]	[2, 10], [10, 2], [6, 10], [10, 6], [3, 9], [5, 4], [3], [4]

The neighborhood of any element of $\mathfrak{C}_2$ - $\mathfrak{C}_1$ is seen to be the cube graph. Note that each of these orthomorphisms must have degree at least 8, and hence, by our computer generated data, they and their translates must have degree exactly 8. Thus we have accounted for all orthomorphisms of degree 8 and their adjacencies and 52 of the adjacencies for each of the orthomorphisms of degree 162. If we restrict ourselves to the orthomorphism graph induced by the elements of $\mathfrak{C}_2$ and their translations then the computer generated data can be completely explained by the theory developed so far. These orthomorphisms also account for all Singer's orbits of length 66 or less.

What of $\mathfrak{C}_5$ - $\mathfrak{C}_1$? The cyclotomy classes are $C_0 = \{1, 10\}$, $C_1 = \{2, 9\}$, $C_2 = \{4, 7\}$, $C_3 = \{3, 8\}$, $C_4 = \{5, 6\}$. It is easily checked that $\sigma_1 = [10, 4, 4, 6, 7]$ and $\sigma_2 = [7, 3, 3, 8, 5]$ are adjacent orthomorphisms of $\mathfrak{C}_5$ and that σ_1 is also adjacent to [5] and $\sigma_4 = [2, 3, 3, 10, 8]$, an orthomorphism of $\mathfrak{C}_5$. Similarly it is easily checked that σ_2 is also adjacent to [4] and $\sigma_3 = [10, 9, 9, 2, 4]$, an orthomorphism of $\mathfrak{C}_5$. It is also easy to see that σ_1 and σ_2 are in different orbits of length 330. This accounts for all vertices of degree 3.

Have we accounted for all of $\mathcal{C}_5$? There are two further orbits of lengths 330 to be explained. Note that the formula of Theorem 3.11 is to unwieldy for calculating $|\mathcal{C}_5 - \mathcal{C}_1|$ as each term in the sum is either 1 or 0, and hence the number of positive terms in the sum is equal to $|\mathcal{C}_5 - \mathcal{C}_1|$. In Singer's list of representatives θ_5 and θ_8 and their images under the congruences T_g are not in $\mathcal{C}_5$ and so we have indeed accounted for all orthomorphisms of $\mathcal{C}_5$. The structure of $\mathrm{Orth}(Z_{11})$ has now been completely described using a mixture of computer generated and theoretically generated data. Its structure is summarized in the following diagram, where vertices of degree 0 are not represented.

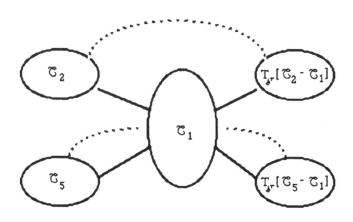

Section 5. Groups of order at least 12.

There are 5 groups of order 12, $Z_{12}, Z_2 \times Z_6, A_4, D_6$, and $T = <a, b|\ a^6 = e, b^2 = a^3, ba = a^{-1}b>$. The groups Z_{12} and T have no orthomorphisms by Theorem 1.20. Chang and Tai (1964) found that $|\mathrm{Orth}(A_4)| = 3776$, $\omega(\mathrm{Orth}(A_4)) = 1$, and $\omega(\mathrm{Orth}(D_6)) \geq 2$. Chang, Hsiang, and Tai (1964) found that $|\mathrm{Orth}(D_6)| = 6336$, and $\omega(\mathrm{Orth}(D_6)) = 2$, and Baumert and Hall (1973) found that $\omega(\mathrm{Orth}(A_4))$ and $\omega(\mathrm{Orth}(D_6))$ were both at most 3.

For the group $Z_6 \times Z_2$, which we will describe as $\{n + ki:\ n \in Z_6, k \in Z_2\}$, Johnson, Dulmage, and Mendelsohn (1961) found the following 7 orthomorphisms.

x	$\theta_1(x)$	$\theta_2(x)$	$\theta_3(x)$	$\theta_4(x)$	$\theta_5(x)$	$\theta_6(x)$	$\theta_7(x)$
0	0	0	0	0	0	0	0
1	i	$4+i$	$5+i$	$5+i$	3	$2+i$	4
2	$2+i$	$1+i$	$4+i$	$4+i$	i	1	$5+i$
3	2	5	$2+i$	5	1	$5+i$	$4+i$
4	$1+i$	$2+i$	2	1	$3+i$	5	2
5	1	3	4	$2+i$	$5+i$	$3+i$	$1+i$
i	$3+i$	2	1	$1+i$	2	3	$2+i$
$1+i$	$5+i$	1	$3+i$	2	$2+i$	$4+i$	i
$2+i$	4	$5+i$	5	i	5	2	$3+i$
$3+i$	$4+i$	4	3	3	4	$1+i$	1
$4+i$	5	$3+i$	$1+i$	$3+i$	$1+i$	i	3
$5+i$	3	i	i	4	$4+i$	4	5

θ_1 is adjacent to each of $\theta_2, \ldots, \theta_7$. θ_2, θ_3, and θ_4 are independent and $\theta_1, \theta_5, \theta_6, \theta_7$ forms a 4 - clique, thus implying the existence of 5 mutually orthogonal Latin squares of order 12. Johnson, Dulmage, and Mendelsohn (1961) cite Parker and Van Duren as having proved via a computer search that $\omega(\text{Orth}(Z_6 \times Z_2)) = 4$. This latter result was later confirmed by Baumert and Hall (1973) again using a computer search.

Bose, Chakravarti, and Knuth (1960) also constructed 4 - cliques of $\text{Orth}(Z_6 \times Z_2)$ using a computer. Their approach was somewhat different. An r-clique of $\text{Orth}(Z_6 \times Z_2)$ corresponds to a $(12, r + 2; 1, Z_6 \times Z_2)$ - difference matrix, which corresponds to a $(12, r + 2; 6, Z_2)$ - difference matrix by Theorem 1.10. Thus, given a difference matrix over Z_2, we might try to reconstruct a difference matrix over $Z_6 \times Z_2$. Any Hadamard matrix of order 12 is a difference matrix over Z_2. Bose, Chakravarti, and Knuth used such a Hadamard matrix and attempted to reconstruct a difference matrix over $Z_6 \times Z_2$ row by row. The cliques that they discovered in the process are listed below.

$\phi_1, \ldots, \phi_4$ constitute a 4 - clique where $\phi_1, \ldots, \phi_4$ are as follows.

x	$\phi_1(x)$	$\phi_2(x)$	$\phi_3(x)$	$\phi_4(x)$
0	0	0	0	0
1	2	$4+i$	5	i
2	1	4	$3+i$	5
3	$4+i$	$2+i$	2	$1+i$
4	$3+i$	2	$4+i$	$5+i$
5	$5+i$	i	$1+i$	4
i	3	$3+i$	$2+i$	2
$1+i$	5	1	4	$2+i$
$2+i$	4	$1+i$	$5+i$	$4+i$
$3+i$	$1+i$	5	1	3
$4+i$	i	$5+i$	3	1
$5+i$	$2+i$	3	i	$3+i$

Also $\psi_1, \ldots, \psi_4$ constitute a 4 - clique where $\psi_1, \ldots, \psi_4$ are as follows.

x	$\psi_1(x)$	$\psi_2(x)$	$\psi_3(x)$	$\psi_4(x)$
0	0	0	0	0
1	2	$1+i$	$2+i$	$3+i$
2	1	4	3	$5+i$
3	$1+i$	$5+i$	2	1
4	i	2	$4+i$	5
5	$2+i$	$3+i$	$1+i$	$4+i$
i	$3+i$	3	5	$2+i$
$1+i$	$5+i$	$4+i$	4	2
$2+i$	$4+i$	1	$5+i$	$1+i$
$3+i$	4	$2+i$	1	i
$4+i$	3	5	i	4
$5+i$	5	i	$3+i$	3

Remarkably, very little information is available for the structure of $\mathrm{Orth}(Z_{13})$. Hsu (1991) reports that Z_{13} has 79,259 orthomorphisms. Cates and Killgrove (1981) reported that the only 11-clique of $\mathrm{Orth}(Z_{13})$ is $\mathfrak{C}_1$. This was confirmed by Mendelsohn and Wolk (1984). These results were obtained through computer searches. Very little theoretical work on $\mathrm{Orth}(Z_{13})$ has been done. We know that Z_{13} has 20 quadratic nonlinear orthomorphisms (listed in Example 3.3), and 54 nonlinear cyclotomic orthomorphisms of index 3 (see Example 3.11).

The only group of order 15 is Z_{15}. Hsu (1991) reports that Z_{15} has 2,424,195 orthomorphisms. It has now been established that $\omega(\mathrm{Orth}(Z_{15})) = 3$. A lower bound of 2 on the clique number had been established by Keedwell (1966) using a computer search. This was improved to 3 by Schellenberg, Van Rees, and Vanstone (1978). Their method was as follows. They tried by computer to construct difference matrices over Z_{15} of the form

$\begin{bmatrix} 0 & O & O \\ O & D & -D \end{bmatrix}$, where the matrices O are zero row or column vectors as appropriate. This is equivalent to searching for a clique of $\text{Orth}(Z_{15})$ consisting of orthomorphisms fixed by H_α, where α is the automorphism $x \to -x$.

They found the following clique.

x	$\psi_1(x)$	$\psi_2(x)$	$\psi_3(x)$
0	0	0	0
1	2	6	10
2	5	3	6
3	7	14	1
4	9	10	11
5	12	7	2
6	4	13	7
7	1	4	12
8	14	11	3
9	11	2	8
10	3	8	13
11	6	5	4
12	8	1	14
13	10	12	9
14	13	9	5

Thus there exist 4 mutually orthogonal Latin squares of order 15. Roth and Wilson have since established that no larger clique of $\text{Orth}(Z_{15})$ can be found, via a computer search. Their result is not yet published. Schellenberg, Van Rees, and Vanstone (1978) also found, using a computer, that $\omega(Z_{33}) \geq 2$, $\omega(Z_{39}) \geq 2$, and $\omega(Z_{21}) \geq 3$.

For the group $Z_2 \times Z_2 \times Z_6$ Roth and Peters (1987) constructed several 3 - cliques using a computer. The orthomorphisms they found are listed in the following table. Here we describe $Z_6 \times Z_2 \times Z_2$ as $\{i\, j\, k: \ i \in Z_6, j \in Z_2, k \in Z_2\}$.

x	$\sigma_1(x)$	$\sigma_2(x)$	$\sigma_3(x)$	$\sigma_4(x)$	$\sigma_5(x)$	$\sigma_6(x)$	$\sigma_7(x)$	$\sigma_8(x)$	$\sigma_9(x)$	$\sigma_{10}(x)$
000	000	000	000	000	000	000	000	000	000	000
100	200	300	210	511	001	510	310	300	311	011
200	400	510	211	010	100	211	111	501	401	101
300	501	201	411	510	101	400	100	211	411	400
400	311	210	501	411	501	301	211	200	301	510
500	410	301	110	001	111	501	501	311	510	501
010	011	001	401	410	400	410	411	001	400	411
110	211	411	010	011	300	310	210	400	200	401
210	511	101	500	110	510	401	201	110	410	200
310	110	400	011	501	301	111	300	401	010	500
410	100	511	301	400	201	011	311	510	201	010
510	301	100	400	211	500	500	400	101	511	001
001	010	011	101	101	110	110	110	011	100	100
101	210	401	310	300	010	010	511	410	500	110
201	510	111	200	401	200	101	500	100	110	511
301	111	410	311	210	011	411	001	411	310	211
401	101	501	001	111	511	311	010	500	501	301
501	300	110	100	500	210	200	101	111	211	310
011	001	010	300	311	310	300	301	010	300	311
111	201	310	510	200	311	210	011	310	011	300
211	401	500	511	301	410	511	410	511	101	410
311	500	211	111	201	411	100	401	201	111	111
411	310	200	201	100	211	001	510	210	001	201
511	411	311	410	310	401	201	200	301	210	210

Of these orthomorphisms the following sets form 3 - cliques: $\{\sigma_1, \sigma_2, \sigma_3\}$, $\{\sigma_1, \sigma_2, \sigma_4\}$, $\{\sigma_1, \sigma_2, \sigma_5\}$, $\{\sigma_1, \sigma_2, \sigma_6\}$, $\{\sigma_1, \sigma_2, \sigma_7\}$, $\{\sigma_1, \sigma_8, \sigma_9\}$, $\{\sigma_1, \sigma_8, \sigma_{10}\}$. This implies the existence of 4 mutually orthogonal Latin squares of order 24. Roberts (Preprint) has improved on this result. He proved the existence of 5 mutually orthogonal Latin squares of order 24 by constructing a 4-clique of $Z_2 \times Z_2 \times Z_6$. Each of Roberts' orthomorphisms is a cyclic permutation of order 23 of the non-zero elements of $Z_2 \times Z_2 \times Z_6$.

Mills (1977) gave a construction of cliques of $\text{Orth}(GF(q)^+ \times GF(p)^+)$, where p is an odd prime divisor of $q - 1$. First, for each $\alpha \in GF(q)$ let $Q_\alpha(x) = a_\alpha x^2 + b_\alpha x + c_\alpha$ be a polynomial over $GF(p)$ with $a_\alpha \neq 0$. Define a p by pq matrix $D(q, p)$ with rows $\{R_g : g \in GF(p)\}$, the columns of each row being indexed by the elements of $GF(q)$, the $\alpha \times i$th entry of R_g being $\alpha \omega^g \times Q_\alpha(i - g)$, where ω is a primitive pth root of unity in $GF(q)$. We define a row vector D_d whose entry in the $\alpha \times i$th column is $\omega^i \beta_{\alpha d} \times (b_{\alpha d} + id)$.

Theorem 6.5. Mills (1977).

i) $D(q, p)$ is a $(qp, p; 1, \mathrm{GF}(q)^+ \times \mathrm{GF}(p)^+)$ - difference matrix.

ii) $\begin{bmatrix} D(q, p) \\ D_d \end{bmatrix}$ is a $(qp, p + 1; 1, \mathrm{GF}(q)^+ \times \mathrm{GF}(p)^+)$ - difference matrix if and only if $\begin{bmatrix} R_0 \\ D_d \end{bmatrix}$ is a $(qp, 2; 1, \mathrm{GF}(q)^+ \times \mathrm{GF}(p)^+)$ - difference matrix.

Proof. i) For $g, k \in \mathrm{GF}(p)$, $g \neq k$, $\alpha\omega^g \times Q_\alpha(i - g) - \alpha\omega^k \times Q_\alpha(i - k) = \beta\omega^g \times Q_\beta(j - g) - \beta\omega^k \times Q_\beta(j - k)$ if and only if $\alpha(\omega^g - \omega^k) = \beta(\omega^g - \omega^k)$ and $Q_\alpha(i - g) - Q_\alpha(i - k) = Q_\beta(j - g) - Q_\beta(j - k)$, if and only if $\alpha = \beta$ and $a_\alpha[(i - g)^2 - (i - k)^2] + b_\alpha[(i - g) - (i - k)] = a_\alpha[(j - g)^2 - (j - k)^2] + b_\alpha[(j - g) - (j - k)]$, if and only if $\alpha = \beta$ and $2ia_\alpha(k - g) = 2ja_\alpha(k - g)$, if and only if $\alpha = \beta$ and $i = j$.

ii) $\omega^i\beta_{\alpha d} \times (b_{\alpha d} + id) - \alpha \times Q_\alpha(i) = \omega^j\beta_{\beta d} \times (b_{\beta d} + jd) - \beta \times Q_\beta(j)$ if and only if $\omega^i\beta_{\alpha d} - \alpha = \omega^j\beta_{\beta d} - \beta$ and $b_{\alpha d} + id - Q_\alpha(i) = b_{\beta d} + jd - Q_\beta(j)$. Also $\omega^i\beta_{\alpha d} \times (b_{\alpha d} + id) - \alpha\omega^g \times Q_\alpha(i - g) = \omega^j\beta_{\beta d} \times (b_{\beta d} + jd) - \beta\omega^g \times Q_\beta(j - g)$ if and only if $\omega^{i - g}\beta_{\alpha d} - \alpha = \omega^{j - g}\beta_{\beta d} - \beta$ and $b_{\alpha d} + (i - g)d - Q_\alpha(i - g) = b_{\beta d} + (j - g)d - Q_\beta(j - g)$. ∎

Using this approach Mills was able to extend the difference matrix $D(13, 3)$ by adding 2 extra rows, thus yielding a construction of 4 mutually orthogonal Latin squares of order 39; $D(13, 3)$ by adding 2 extra rows, thus yielding a construction of 4 mutually orthogonal Latin squares of order 39; $D(16, 3)$ by adding 2 extra rows, thus yielding a construction of 4 mutually orthogonal Latin squares of order 48; $D(11, 5)$ by adding 1 extra row, thus yielding a construction of 5 mutually orthogonal Latin squares of order 55; $D(8, 7)$ by adding 1 extra row, thus yielding a construction of 7 mutually orthogonal Latin squares of order 56; $D(4, 3)$ by adding 3 extra rows, thus giving another construction of 5 mutually orthogonal Latin squares of order 12. The following is the difference matrix he obtained by extending $D(4, 3)$, which we are writing in transposed form.

0, 0	0, 0	0, 2	1, 0	1, 0	$\omega^2, 0$
0, 2	0, 0	0, 0	$\omega, 0$	$\omega, 1$	1, 2
0, 0	0, 2	0, 0	$\omega^2, 0$	$\omega^2, 2$	$\omega, 1$
1, 2	$\omega, 1$	$\omega^2, 2$	1, 0	$\omega^2, 2$	$\omega, 0$
1, 2	$\omega, 2$	$\omega^2, 1$	$\omega, 0$	1, 0	$\omega^2, 2$
1, 1	$\omega, 2$	$\omega^2, 2$	$\omega^2, 0$	$\omega, 1$	1, 1
$\omega, 1$	$\omega^2, 2$	1, 1	1, 0	$\omega, 1$	1, 0
$\omega, 1$	$\omega^2, 1$	1, 2	$\omega, 0$	$\omega^2, 2$	$\omega, 2$
$\omega, 2$	$\omega^2, 1$	1, 1	$\omega^2, 0$	1, 0	$\omega^2, 1$
$\omega^2, 0$	1, 0	$\omega, 1$	1, 0	$\omega, 2$	$\omega, 1$
$\omega^2, 1$	1, 0	$\omega, 0$	$\omega, 0$	$\omega^2, 0$	$\omega^2, 0$
$\omega^2, 0$	1, 1	$\omega, 0$	$\omega^2, 0$	1, 1	1, 2

Chapter 7: Research directions

Section 1. Problems to consider.

In Chapter 1, Section 1, the following general problems were listed. Given a group G and an orthomorphism graph $\mathcal{H}$ of G:

i) Determine exact values of, or bounds for, $\omega(\mathcal{H})$.
ii) What can we say about the structure of $\mathcal{H}$?
iii) Can $\mathcal{H}$ contain a complete set of orthomorphisms?
iv) Given a clique of $\mathcal{H}$ can we extend it to a larger clique of $\mathcal{H}$ or $\mathrm{Orth}(G)$?

We have also considered the following special cases of these problems.

v) Is $\mathcal{H}$ non-empty?
vi) Does there exist an r-clique in $\mathcal{H}$ that is maximal in $\mathcal{H}$ or $\mathrm{Orth}(G)$?

Answering problem v) is equivalent to answering the question "Is $\omega(\mathcal{H}) \geq 1$?", and in answering problem vi) we are essentially looking for r-cliques of $\mathcal{H}$ for which the answer to problem iv) is negative.

In this chapter we will discuss the progress made toward solving these and related problems, as well as possible future directions for research, and we will suggest several specific problems.

These discussions and problems are loosely organized into three sections. In Section 2 we will discuss known orthomorphism graphs, and the search for other interesting orthomorphism graphs. Section 3 will be concerned with the structure of orthomorphism graphs, both the global structure, that is the structure of $\mathrm{Orth}(G)$, and the local structure, that is the structure of individual orthomorphism graphs such as $\mathcal{P}(G)$. Section 4 deals with cliques, in particular bounds on clique numbers, maximal cliques, complete sets of orthomorphisms, and the existence of orthomorphisms and complete mappings, i.e. 1-cliques.

Section 2. Orthomorphism graphs.

While the study of orthomorphism graphs is of interest in its own right, the original impetus for their study comes from attempts to construct large sets of mutually orthogonal Latin squares, and complete sets of mutually orthogonal Latin squares. More recently orthomorphism graphs have been used in constructions of maximal sets of mutually orthogonal

Latin squares and in the study of orthogonal Latin square graphs. There are also several reasons to study individual orthomorphisms and their constructions: In Chapter 1, Section 4 we saw their use in the construction of neofields. Other applications of individual orthomorphisms can be found in the papers reprinted in Hsu (1985 and 1987).

At present it would appear that our primary interest in the study of orthomorphism graphs lies with the study of those orthomorphism graphs that contain large cliques, maximal cliques of the orthomorphism graph, complete sets of orthomorphisms, or significant pieces of such cliques. The orthomorphism graphs that we have studied the most are those of elementary abelian groups. What is the rationale for this? Large sets of mutually orthogonal Latin squares are only seen to be of interest for non-prime-power order. Complete sets of orthomorphisms of elementary abelian groups are generally studied by other means (see the extensive literature on translation planes). However, the search for complete sets of orthomorphisms of Z_p, p a prime, is of interest. The goal of this search is to find a non-Desarguesian plane of prime order, or to prove that Cartesian planes of prime order must be Desarguesian. This is a special case of the more general conjecture that all planes of prime order must be Desarguesian. This was the original justification for the study of orthomorphism graphs of cyclic groups of prime order, and the techniques used and the results proved generalized to elementary abelian groups with only a little extra work. It is reasonable to expect many maximal cliques to be discovered in orthomorphism graphs of elementary abelian groups, even though the only large class of such cliques we have seen (see Theorem 5.16) lies in $\mathfrak{C}_2(p)$, p a prime, and its translates. Other maximal cliques have been discovered: Many constructions of maximal translation nets in, for example, Beth, Jungnickel, and Lenz (1985, Chapter X) are equivalent to constructions of maximal cliques of $\mathfrak{A}(G)$, $G = GF(q)^+$. However, orthomorphisms are not used in these constructions.

This falsely suggests that the study of orthomorphism graphs of elementary abelian groups is only of interest in constructions of maximal cliques and in the search for non-Desarguesian planes of prime order. Actually there are other reasons that we can advance for the study of these graphs. There are far more theoretical tools available for studying combinatorial problems in elementary abelian groups than in groups in general: We have used field theory, linear algebra, cyclotomy, and permutation polynomials. Thus elementary abelian groups form an ideal environment for developing constructions of orthomorphisms and for testing methods for studying orthomorphism graphs, which we might then try to adapt to the study of orthomorphism graphs of groups in general. Cyclic groups of prime order form the basic building blocks for abelian groups and solvable groups. Thus we might expect orthomorphism graphs of cyclic groups of prime order to be useful building blocks in the construction of orthomorphism graphs of abelian and solvable groups. New ways to construct orthomorphisms of $G \times H$ from orthomorphisms of the groups G and H could yield new interesting orthomorphism graphs. Of particular interest is the case in which G and H are elementary abelian groups. We have seen several instances of large sets of mutually orthogonal Latin squares being constructed from orthomorphisms of direct products of elementary abelian groups: 5 mutually orthogonal Latin squares of order 12 based on the group $(Z_2 \times Z_2) \times Z_3$; 4 mutually orthogonal Latin squares of order 15 based on the group $Z_3 \times Z_5$; and 5 mutually orthogonal Latin squares of order 24 based on the group $(Z_2 \times Z_2 \times Z_2) \times Z_3$, to name but a few.

In this section we will be interested in finding new, interesting orthomorphism graphs, and new ways to combine orthomorphism graphs. Given a group G and a normal subgroup H of G we would like to find new ways to construct orthomorphisms of G from orthomorphisms of G/H and H and so combine orthomorphism graphs of G/H and H to form new orthomorphism graphs of G. Closely related to this is the problem of finding new ways to extend orthomorphism graphs. Given a group G and a normal subgroup H of G we would like to find new ways to extend an orthomorphism of G/H or H to an orthomorphism of G and so extend orthomorphism graphs of G/H or H to form new orthomorphism graphs of G.

What should we ask of any new orthomorphism graph? Rules for membership and adjacency should be as manageable as possible. As an example, although $\mathfrak{C}_2(q)$ and $T_g[\mathfrak{C}_2(q)]$ are isomorphic orthomorphism graphs, the membership and adjacency rules are clearly much simpler for $\mathfrak{C}_2(q)$. We should have reason to believe that our new orthomorphism graph forms an important part of Orth(G). At the very least this new graph should contain edges. We should prefer it to contain reasonably large cliques, extendable cliques, or cliques that are maximal in Orth(G).

In Chapter 2 we constructed orthomorphisms of $GF(q)^+$ by solving systems of equations of the form $ax_{\sigma(i)} - x_i = (a - 1)x_{\varepsilon(i)}$, σ and ε permutations of $\{1, \ldots, q - 1\}$, for $a, x_1, \ldots, x_{q-1}$. These systems of equations we called (σ, ε)-systems. This approach has its advantages and disadvantages. It is a good method for generating classes of orthomorphisms, and so orthomorphism graphs, fairly quickly. However this approach does not yield much information about the structure of the generated orthomorphism graphs: For this we must use other methods. Since the number of (σ, ε)-systems increases dramatically as q increases this is not a practical method for calculating all orthomorphisms for $GF(q)^+$ for even fairly small values of q. To use this approach to calculate all orthomorphisms of $GF(q)^+$ we need to solve, up to equivalence, all (σ, ε)-systems of order $q - 3$ or less. For $q = 9$ this means solving 36 inequivalent systems of order 6, those of order 5 or less having been solved in Chapter 2.

Here are some ideas for the further study of (σ, ε)-systems.

Problem 1. Can Theorem 2.4 be strengthened? This theorem concerns the (σ, ε)-system for $GF(q)$ with $\sigma = (1\ 2\ \ldots\ n)$ and $\varepsilon = \sigma^r$, $2 \le r < n \le q - 1$. Is it true that this system yields orthomorphisms if and only if n divides $q - 1$? If n divides $q - 1$ are the orthomorphisms yielded by this system precisely those described in Theorem 2.3? Can the rank of the solution space of this (σ, ε)-system ever be greater than 2?

Problem 2. Consider again the (σ, ε)-system with $\sigma = (1\ 2\ \ldots\ n)$ and $\varepsilon = \sigma^r$, $2 \le r < n$, and let R be a finite commutative ring with unity. Under what conditions will this system yield orthomorphisms of R^+? Describe the orthomorphisms yielded by this system. In Chapter 2, when studying this type of (σ, ε)-system, we restricted ourselves to additive groups of finite fields.

Problem 3. Find new interesting classes of (σ, ε)-systems, determine the orthomorphisms yielded by these systems, and study the orthomorphism graphs these determine over finite fields or rings.

We might, for instance, define new classes by splicing together (σ, ε)-systems. As an example, the system defined by $\sigma = (1\ 2\ 3)$ and $\varepsilon = (1\ 3\ 2)$ could be spliced with itself to define a new system with $\sigma' = (1\ 2\ 3)(4\ 5\ 6)$ and $\varepsilon' = (1\ 3\ 2)(4\ 6\ 5)$. The solutions to the new system can be obtained from solutions to the original system.

In studying orthomorphism graphs of $GF(q)^+$ we used (σ, ε)-systems, cyclotomy, and permutation polynomials. Theorem 3.16 characterizes permutation polynomials that represent cyclotomic orthomorphisms. However we have no similar theorems relating (σ, ε)-systems to cyclotomic orthomorphisms and permutation polynomials. Problems 4, 5, and 6 ask for ways to address this deficiency.

Problem 4. Can we determine a (σ, ε)-system that yields $H_\alpha[\theta]$ from a (σ, ε)-system that yields θ? More generally, if F is a congruence of $Orth(GF(q)^+)$, can we determine a (σ, ε)-system that yields $F[\theta]$ from a (σ, ε)-system that yields θ?

Problem 5. Given $e \mid q - 1$, characterize those (σ, ε)-systems of $GF(q)$ that yield cyclotomic orthomorphisms of $GF(q)^+$ of index e.

Problem 6. Given a (σ, ε)-system of $GF(q)$, characterize those permutation polynomials of $GF(q)$ that represent orthomorphisms of $GF(q)^+$ yielded by this system.

Problem 7. Suppose R to be a commutative ring with unity and let $a \in R$, with a and $a - 1$ both units of R. If θ is an orthomorphism of $R^+ = \{r_1, \dots, r_n\}$ then $\theta(r_i) = ar_{\sigma(i)}$ and $\theta(r_i) - r_i = (a - 1)r_{\varepsilon(i)}$, for some σ and ε permutations of $\{1, \dots, n\}$. This defines a (σ, ε)-system. Use this approach to construct new (σ, ε)-systems, and hence new orthomorphism graphs, from known cliques of orthomorphism graphs of abelian groups.

Problem 8. Is it useful to talk about (σ, ε)-systems for groups in general? Let $G = \{g_1, \dots, g_n\}$ be a finite group. Any orthomorphism of a group G can be written in the form $\theta(g_i) = \alpha(g_{\sigma(i)})$, for some $\alpha \in Aut(G)$ and some permutation σ of $\{1, \dots, n\}$, and $g_i^{-1}\theta(g_i) = g_i^{-1}\alpha(g_{\sigma(i)}) = \beta(g_{\varepsilon(i)})$, for some $\beta \in Aut(G)$ and some permutation ε of $\{1, \dots, n\}$. Thus we might try to solve systems of equations of the form $x_i^{-1}\alpha(x_{\sigma(i)}) = \beta(x_{\varepsilon(i)})$, $\alpha, \beta \in Aut(G)$, σ, ε permutations of $\{1, \dots, n\}$. If G is abelian and α is a fixed point free automorphism of G then we might set $\beta(g) = g^{-1}\alpha(g)$ as this will also be an automorphism of G: We have already seen the special case $G = R^+$, R a commutative ring with unity, $\alpha(x) = ax$, a and $a - 1$ both units of R.

In Chapter 3 we introduced quadratic and cyclotomic orthomorphisms. In contrast to (σ, ε)-systems, which yield large classes of orthomorphisms with relative ease, but offer little information on adjacencies, the use of cyclotomy yields simultaneously constructions of orthomorphisms and simple rules for adjacency between these orthomorphisms. In addition cyclotomic orthomorphisms can be viewed in several different ways, allowing several generalizations to arbitrary groups. For g a primitive element of GF(q), $q = ef + 1$, the ith cyclotomy class is $C_i(q) = \{g^{i + ej}: j = 0, \ldots, f - 1\}$, $i = 0, \ldots, e - 1$. There are three different but equivalent ways to view cyclotomic orthomorphisms. They are orthomorphisms of the form $[A_0, \ldots, A_{e-1}]$: $x \to A_i x$ if $x \in C_i$, and 0 if $x = 0$, where $[A_0, \ldots, A_{e-1}]$ is an orthomorphism if both $\{A_0 g^0, \ldots, A_{e-1} g^{e-1}\}$ and $\{(A_0 - 1)g^0, \ldots, (A_{e-1} - 1)g^{e-1}\}$ are systems of distinct representatives for the family of cyclotomy classes. Another approach: Let $K = \{x \to ax: a \in C_0\}$. Then an orthomorphism θ is a cyclotomic orthomorphism if and only if $H_\alpha[\theta] = \theta$ for all $\alpha \in K$. Note that the cyclotomy classes are the orbits of K. Lastly the set of cyclotomic orthomorphisms is represented by the set of orthomorphism polynomials of the form $f(x) = \sum a_i x^{1 + if}$.

Problem 9. Let $\{X_1, \ldots, X_m\}$ be a partition of the set of non-identity elements of a group G. Study classes of orthomorphisms θ of G for which $\theta(X_i) = X_{\sigma(i)}$, for some permutation σ of $\{1, \ldots, m\}$, and the corresponding orthomorphism graph.

Partitions have already been used in the study of orthomorphism graphs. If $K \subseteq$ Aut(G) then the orbits of K form a partition of the elements of G. Cyclotomic orthomorphisms preserve a partition of this type. In Theorem 1.23 we constructed, from orthomorphisms of $H \subseteq G$, orthomorphisms of G that permute the cosets of H. Theorems 1.28, 1.29, and 1.30, establishing the existence of orthomorphisms for certain groups, were all proved using appropriate partitions.

Problem 10. Let K be a group of congruences of Orth(G). Study the orthomorphism graph $\{\theta \in$ Orth(G): $T[\theta] = \theta$ for all $T \in K\}$.

Inherent in our classification of orthomorphisms of GF$(q)^+$ is the notion of classifying orthomorphisms and their corresponding orthomorphism graphs by the congruences that fix them. Cyclotomic orthomorphisms are characterized by the homologies that fix them. Elements of $\mathcal{C}(G)$ are fixed by each element of $\{T_g: g \in G\}$: This characterizes elements of $\mathcal{C}(G)$.

Problem 11. In studying orthomorphisms of $Z_p \times \ldots \times Z_p$ we have used the fact that this group can be thought of as the additive group of a finite field, and have used field multiplication to define linear, quadratic, and cyclotomic orthomorphisms. There are many ways to define multiplication on $Z_p \times \ldots \times Z_p$ so that the field axioms are satisfied. Extend the class of linear (respectively quadratic, or cyclotomic) orthomorphisms to include orthomorphisms that are linear (respectively quadratic, or cyclotomic) with respect to some field multiplication, and study the corresponding orthomorphism graphs. As an example, every orthomorphism of $Z_2 \times Z_2 \times Z_2$ is linear with respect to some field multiplication.

Problem 12. Let R be a ring and let $P(R) = \{f: f \in R(x), r \to f(r)$ is a permutation of $R\}$, the set of permutation polynomials of R. Study the orthomorphism graph $\mathrm{Orth}(R^+) \cap P(R)$. In general not all permutations of the elements of R can be represented by permutation polynomials of R.

Problem 13. Find more classes of permutation polynomials and orthomorphism polynomials.

Problem 14. If q is even, $q \neq 2$, then $\mathrm{Orth}(\mathrm{GF}(q)^*) \neq \varnothing$, and any orthomorphism of $\mathrm{GF}(q)^*$ can be represented by a permutation polynomial of $\mathrm{GF}(q)$ with constant term 0. Use permutation polynomials to study $\mathrm{Orth}(\mathrm{GF}(q)^*)$. The first case of interest is $\mathrm{Orth}(\mathrm{GF}(16)^*) \cong \mathrm{Orth}(Z_{15})$.

Problem 15. Find new interesting orthomorphism graphs by studying subgraphs of the orthomorphism graphs $\mathcal{D}_D(G)$ and $\mathcal{J}_{D,H}(G)$. As an example, if K is a group of order $q + 1$ then each orthomorphism graph of K can be used to define a subgraph of the orthomorphism graph $\mathcal{D}_D(Z_n)$, $n = q^2 + q + 1$.

Problem 16. Theorem 6.5 is stated in difference matrix form. Study the new orthomorphism graph obtained when this theorem is rewritten in orthomorphism form.

Problem 17. In Chapter 5, Section 1 we introduced the mapping $\phi_r(x) = x^r$, and in Section 2 of that chapter we defined a strong complete mapping to be an orthomorphism that is adjacent to ϕ_{-1}. Study the orthomorphism graphs consisting of those orthomorphisms of G that are adjacent to each element of S, $S \subseteq \mathcal{P}(G)$.

Now to the problem of combining orthomorphisms and hence orthomorphism graphs to construct new orthomorphism graphs. We can obtain new orthomorphism graphs by taking images under congruences, e.g. $T_g[\mathcal{C}_2(q)]$, and through using the set theoretic operations of union, intersection, and difference. We have also seen a direct product construction: If $\mathcal{H}$ is an orthomorphism graph of H and $\mathcal{K}$ is an orthomorphism graph of K then $\mathcal{H} \times \mathcal{K} = \{\theta \times \phi: \theta \in \mathcal{H}, \phi \in \mathcal{K}\}$ is an orthomorphism graph of $H \times K$.

Problem 18. If H is a normal subgroup of G find new ways to extend orthomorphisms of H or G/H to orthomorphisms of G, and hence extend orthomorphism graphs of H or G/H to orthomorphism graphs of G. More generally if H is a (not necessarily normal) subgroup of G, find new ways to extend orthomorphisms of H to orthomorphisms of G.

Section 3. Structure of orthomorphism graphs.

In Section 2 we considered the search for new orthomorphism graphs. Having found a new orthomorphism graph we would like to see how it fits in with those orthomorphism graphs already discovered and also what is contained inside our newly discovered orthomorphism graph. In other words we are interested in structure results. Given a group G and a collection of orthomorphism graphs of G, how do these fit together? Given an orthomorphism graph $\mathcal{H}$ of a group G what is the structure of $\mathcal{H}$? The first of these questions deals with global structure, whereas the second deals with local structure.

The following are the general type of questions we might ask when studying global structure.

i)   Given $\mathcal{H}$ and $\mathcal{K}$ orthomorphism graphs of a group G, is any vertex of $\mathcal{H}$ adjacent to any vertex of $\mathcal{K}$?

ii)   Given $\mathcal{H}$ and $\mathcal{K}$ orthomorphism graphs of a group G, how many vertices of $\mathcal{H}$ can a vertex of $\mathcal{K}$ be adjacent to?

iii)   Given $\mathcal{H}$ and $\mathcal{K}$ orthomorphism graphs of a group G, describe $\mathcal{H} \cap \mathcal{K}$.

iv)   Given a family $\{\mathcal{H}_i\}$ of orthomorphisms graphs of G, how much of the structure of $\mathrm{Orth}(G)$ is contained in $\cup \mathcal{H}_i$?

In answering iv), ideally we would like each orthomorphism graph in the family $\{\mathcal{H}_i\}$ to have a "manageable" structure, and $\mathrm{Orth}(G) - \cup \mathcal{H}_i$ to consist of orthomorphisms of degree 0. We are a long way from achieving this ideal. As an example, if $G = \mathrm{GF}(q)^+$ then $\mathrm{Orth}(G) - \cup \{\mathcal{C}_e(q):\ e$ divides $q - 1\}$ is empty as $\mathcal{C}_{q-1}(q) = \mathrm{Orth}(G)$. This is not a very interesting choice of a family of orthomorphism graphs. A better choice would be $\cup \mathcal{H}_i = \cup \{T_g[\mathcal{C}_e(q)]:\ g \in \mathrm{GF}(q), 1 \le e < q - 1, e$ divides $q - 1\}$. With this choice $\mathrm{Orth}(G) - \cup \mathcal{H}_i$ consists of orthomorphisms of degree 0 for $q \le 11$, q a prime. Whether this holds true for all prime q is unknown.

Problem 19. We know that $\mathrm{Orth}(Z_2 \times Z_4)$ consists of 12 disjoint 4-cycles. Give a theoretical proof of this. Does this give us any insight into the structure of $\mathrm{Orth}(Z_2 \times Z_8)$ or more generally $\mathrm{Orth}(Z_2 \times Z_q)$, $q = 2^n$?

Problem 20. Explain the structure of $\mathrm{Orth}(\mathrm{GF}(9)^+)$. The description given in Theorem 6.4 is incomplete.

Problem 21. Explain, without the use of computer generated data, the structure of $\mathrm{Orth}(Z_{11})$.

Problem 21 can be broken down into several smaller problems. Why is no element of $\mathfrak{C}_2(11)$ - $\mathfrak{C}_1(11)$ adjacent to any orthomorphism outside of $\mathfrak{C}_2(11)$? Can we extend this explanation to obtain information about possible adjacencies between elements of $\mathfrak{C}_2(p)$ - $\mathfrak{C}_1(p)$ and orthomorphisms outside of $\mathfrak{C}_2(p)$, p a prime? Similarly, why is no element of $\mathfrak{C}_5(11)$ - $\mathfrak{C}_1(11)$ adjacent to any orthomorphism outside of $\mathfrak{C}_5(11)$? Can we extend this explanation to obtain information about possible adjacencies between elements of $\mathfrak{C}_e(p)$ - $\mathfrak{C}_1(p)$ and orthomorphisms outside of $\mathfrak{C}_e(p)$, p a prime? Prove theoretically that Orth(Z_{11}) has only one 9-clique. Prove theoretically that all edges of Orth(Z_{11}) are within $\mathfrak{C}_1(11)$, $\mathfrak{C}_2(11)$, $\mathfrak{C}_5(11)$, and their translates. Why is each element of $\mathfrak{C}_5(11)$ - $\mathfrak{C}_1(11)$ adjacent to exactly 1 element of $\mathfrak{C}_1(11)$? Can elements of $\mathfrak{C}_e(p)$ - $\mathfrak{C}_1(p)$, p a prime, $e \neq p - 1$, be characterized by the number of elements of $\mathfrak{C}_1(p)$ they are adjacent to ?

Problem 22. Give a theoretical proof that $|\text{Orth}(A_4)| = 3776$ and $|\text{Orth}(D_6)| = 6336$.

Problem 23. Determine the structure of Orth($Z_2 \times Z_6$).

Problem 24. Determine the structure of Orth(Z_{13}).

It would also be interesting to know the structures of the orthomorphism graphs of groups of orders 16, 17, 19, 20, 21, 23, and 24.

Problem 25. Theoretically determine the structure of Orth(Z_{15}).

Problem 26. For p an odd prime, what types of orthomorphisms are allowed as neighbors of a non-linear quadratic orthomorphism of GF$(p)^+$? For $p \leq 11$ we know that only quadratic orthomorphisms are allowed.

Problem 27. For p an odd prime, what types of orthomorphisms are allowed as neighbors of a non-linear cyclotomic orthomorphism of index e, $1 < e < p - 1$? For $p \leq 11$ we know that only cyclotomic orthomorphisms of the same index are allowed.

Problem 28. If $a, b \mid q - 1$ and $a \mid b$ we know that $\mathfrak{C}_a(q) \subseteq \mathfrak{C}_b(q)$. When do we have equality? When do we have inequality?

Problem 29. Theorem 3.11 yields a formula for $|\mathfrak{C}_e(q)|$ which is only practical for small values of e. Find a better formula.

Problem 30. If q is not prime describe $T_g[\mathfrak{C}_a] \cap T_h[\mathfrak{C}_b]$, where $a, b \mid q - 1$, $a, b \neq q - 1$, and $g, h \in \text{GF}(q)$, $g \neq h$. If q is prime then $T_g[\mathfrak{C}_a] \cap T_h[\mathfrak{C}_b] = \mathfrak{C}_1$.

In studying local structure we are concerned with the structure of some given orthomorphism graph or family of orthomorphism graphs. If $\mathcal{H}$ is an orthomorphism graph of a group G then the study of the structure of $\mathcal{H}$ requires us to deal with the following problems.

i)   Characterize the elements of $\mathcal{H}$.

ii)   Determine adjacency rules for elements of $\mathcal{H}$.

iii)   Determine the structure of $\mathcal{H}$.

iv)   Determine the structure of some subgraph of $\mathcal{H}$, such as the neighborhood of an element of $\mathcal{H}$.

The following problems, dealing with cliques, also concern local structure but will be discussed in Section 4 of this chapter.

v)   Determine exact values of, or bounds on, $\omega(\mathcal{H})$.

vi)   Find maximal cliques of $\mathcal{H}$.

vii)   Does $\mathcal{H}$ contain a complete set of orthomorphisms? If so then characterize the complete sets of orthomorphisms contained in $\mathcal{H}$.

The structure of an orthomorphism graph $\mathcal{H}$ is sometimes better understood when $\mathcal{H}$ is described as a subgraph of a larger, better known graph. As an example, we can embed $\mathcal{C}_2(q)$ in the graph Γ, whose vertices are ordered pairs of elements of GF(q), (x, y) being adjacent to (x', y') if $(x - x')(y - y')$ is a non-zero square. Γ is a strongly regular graph being the line graph of the net with points (x, y), $x, y \in$ GF(q), and lines $y = mx + b$, m a non-zero square. $\mathcal{C}_2(q)$ is isomorphic to the subgraph of Γ induced by the vertices of Γ that are adjacent to both $(0, 0)$ and $(1, 1)$, the isomorphism being given by the mapping $[A, B] \to (A, B)$.

In a similarly manner we can describe the structure of $\mathcal{P}(G)$, first noting that $\mathcal{P}(G) \cong \mathcal{P}(Z_{|G|})$. Let $p_1, \ldots, p_m$ be the distinct prime divisors of $|G|$, and define Γ to be the graph whose vertices are the m-tuples $(a_1, \ldots, a_m)$, $0 \le a_i < p_i$ for all i, $(a_1, \ldots, a_m)$ being adjacent to $(b_1, \ldots, b_m)$ if and only if $a_i \ne b_i$ for all i. Define $\eta: \mathcal{P}(G) \to \Gamma$ by $\eta(\phi_r) = (r \bmod p_1, \ldots, r \bmod p_m)$. η maps $\mathcal{P}(G)$ onto the common neighborhood of $(0, \ldots, 0)$ and $(1, \ldots, 1)$ in Γ, and preserves adjacency. If $|G|$ is square-free then η is an isomorphism.

Problem 31. If $[A, B]$ is a non-linear quadratic orthomorphism of GF(q)$^+$ what can we say about the structure of the orthomorphism graph induced by its neighbors in $\mathcal{C}_2(q)$?

Problem 32. If θ is a non-linear cyclotomic orthomorphism of GF(q)$^+$ of index e what can we say about the structure of the orthomorphism graph induced by its neighbors in $\mathfrak{C}_e(q)$?

Problem 33. Theorem 3.19 gives rules for membership and adjacency in the set of cyclotomic orthomorphism polynomials discovered by Niederreiter and Robinson. What can we say about the structure of the orthomorphism graph induced by this set of cyclotomic orthomorphisms?

Problem 34. Determine the structure of $\mathcal{Q}(G)$, $\mathcal{Q}^-(G)$, or $\mathcal{Q}(G) \cup \mathcal{Q}^-(G)$ for some class of groups other than the elementary abelian groups.

Problem 35. Given a graph G determine $r(G)$, the representation number of G, which is defined to be the least n for which G can be represented modulo n.

Two easy examples: The representation number of the complete graph on m vertices is $\{p: p \geq m, p$ a prime$\}$ and the representation number of the edgeless graph on m vertices is $2m$. Bounds on the representation numbers of paths and cycles can be found in Evans, Fricke, Maneri, McKee, and Perkel (Preprint).

Problem 36. Characterize those graphs that can be represented modulo pqr, for some p, q, and r distinct primes.

Graphs that can be represented modulo a prime are complete graphs. Graphs that can be represented modulo a power of a prime are complete multipartite graphs, and graphs that can be represented modulo the product of two distinct primes are characterized by the lack of an induced subgraph isomorphic to $K_2 \cup 2K_1$, $K_3 \cup K_1$, or the complement of an odd chordless cycle of length at least 5. The proofs are in Evans, Fricke, Maneri, McKee, and Perkel (Preprint).

Section 4. Cliques of orthomorphism graphs.

Most of the problems, described in Section 1 of this chapter, concerned cliques of orthomorphism graphs. These problems dealt with the determination of exact values of clique numbers, bounds on clique numbers, the existence of complete sets of orthomorphisms, maximal cliques, and the existence of orthomorphisms. Most of the problems that we will suggest fall into one of the following categories, where G is a given a group.

a)   Determine $\omega(\mathrm{Orth}(G))$.

b)   If by computer it has been determined that $\omega(\mathrm{Orth}(G)) = r$ say, then prove this theoretically.

c)   Construct larger cliques of Orth(G), i.e. improve known lower bounds for ω(Orth(G)).

d)   Find new maximal cliques of Orth(G).

e)   If we have found a new maximal clique of Orth(G) by computer, then give a theoretical prove of its maximality.

f)   Given a maximal clique of an orthomorphism graph $\mathcal{H}$ of G, decide if it is maximal in Orth(G) or not.

g)   Can we add G to our list of groups known not to admit complete sets of orthomorphisms?

h)   If G = GF(q)$^+$ and $\mathcal{H}$ is an orthomorphism graph of G then characterize the possible complete sets of orthomorphisms contained in $\mathcal{H}$.

i)   Does G admit orthomorphisms, strong complete mappings, etc?

The following list gives more specific problems.

Problem 37. Prove theoretically that ω(Orth(D_4)) = 1.

Problem 38. Determine ω(Orth(D_8)). What can we say about ω(Orth(D_q)), $q = 2^n$, or ω(Orth(D_{2n}))?

Problem 39. Prove theoretically that ω(Orth(Q_8)) = 1.

Problem 40. Prove theoretically that ω(Orth(Z_9)) = 1.

Problem 41. Determine ω(Orth(Z_{27})).

Problem 42. We know that, if $q = p^n$, p a prime then ω(Orth(Z_q)) $\geq p$ - 2. Do we have equality? Is there a value of q for which ω(Orth(Z_q)) $> p$ - 2?

Problem 43. Prove theoretically that ω(Orth(A_4)) = 1.

Problem 44. Prove theoretically that ω(Orth($Z_2 \times Z_6$)) = 4.

Problem 45. Prove theoretically that ω(Orth(Z_{15})) = 3.

Problem 46. Determine clique numbers for the orthomorphism graphs of groups of orders 16, 17, 19, 20, 21, 23, and 24.

Problem 47. If $p < q$ are odd primes we know that $\omega(\text{Orth}(Z_{pq})) \geq p - 2$. This is the best available lower bound for most values of p and q. Try to improve this bound.

Problem 48. Given a λ-difference matrix over H and epimorphism $G \rightarrow H$, $|G|/|H| = \lambda$ try to construct large cliques of $\text{Orth}(G)$. Bose, Chakravarti, and Knuth (1960) used this approach with $H = Z_2$ and $\lambda = 6$.

Problem 49. If $|G|$ is odd, when is $\omega(\mathcal{B}(G)) = \omega(\text{Orth}(G)) - 1$? When is $\omega(\mathcal{B}(G)) < \omega(\text{Orth}(G)) - 1$? If $|G|$ is even, when is $\omega(\mathcal{B}(G)) = \omega(\text{Orth}(G))$? When is $\omega(\mathcal{B}(G)) < \omega(\text{Orth}(G))$?

Problem 50. In Theorem 4.3 we proved that $\omega(A_{\mathcal{H}} \cup \text{Orth}(G)) \leq \omega(A_H \cup \text{Orth}(G))$, where $A_{\mathcal{H}}$ consists of those elements of $\mathcal{C}(G)$ and $\mathcal{C}^-(G)$ that permute the conjugates of H, and A_H consists of those elements of $\mathcal{C}(G)$ and $\mathcal{C}^-(G)$ that fix H. When does equality hold?

Problem 51. In Corollary 4.9 we proved that $\omega(\mathcal{C}(G)) \leq \text{Min}\{\omega(\mathcal{C}(H)), \omega(\mathcal{C}(G/H))\}$ if H char G. When does equality hold?

Problem 52. In Corollary 4.10 we proved that $\omega(\mathcal{C}(G \times H)) = \text{Min}\{\omega(\mathcal{C}(H)), \omega(\mathcal{C}(G))\}$ if $|G|$ and $|H|$ are relatively prime. What happens if $|G|$ and $|H|$ are not relatively prime?

Problem 53. Determine $\omega(\mathcal{C}(G))$ for (a class of) non-abelian p-groups. Some upper bounds can be found in Bailey and Jungnickel (1990).

Problem 54. For what values of n is $R(n) < N(n)$? For what values of n is $R(n) = N(n)$?

If $n \equiv 2$ modulo 4 and $n > 6$ then $R(n) < N(n)$. It is not known if $R(n) < N(n)$ for any other value of $n > 6$. Of course $R(n) = N(n)$ if n is a power of a prime.

Problem 55. Let L be a Cayley table for a group G and let $\theta_1, \ldots, \theta_s$ be a clique of $\text{Orth}(G)$. If $L, L(\theta_1), \ldots, L(\theta_s), M_{s+1}, \ldots, M_r$ is a mutually orthogonal set of Latin squares can $\theta_1, \ldots, \theta_s$ be extended to an r-clique of $\text{Orth}(G)$?

Problem 56. Let p be the smallest prime divisor of $|G|$ and let S be a Sylow p-subgroup of G. We know that $\omega(\mathcal{P}(G)) = p - 2$ and the only maximal cliques of $\mathcal{P}(G)$ are $(p-2)$-cliques. When are maximal cliques of $\mathcal{P}(G)$ also maximal in $\mathcal{P}(G)$?

If S is cyclic then there exists a homomorphism from G onto S and Corollary 5.14 then implies that maximal cliques of $\mathcal{P}(G)$ are maximal in Orth(G). Thus this question is only of interest when S is noncyclic. On the other hand if $G = S = GF(q)^+, q > p$, then no maximal clique of $\mathcal{P}(G)$ is maximal in Orth(G). It is also of interest to know, when these cliques are not maximal, how far they can be extended.

Problem 57. $[10, 4, 4, 6, 7] \in \mathfrak{C}_5(11)$ and $[5]$ together form a maximal 2-clique of Orth(Z_{11}). Prove this theoretically and so obtain a theoretical proof that there exists a maximal set of 3 mutually orthogonal Latin squares of order 11. If possible generalize your argument.

Problem 58. For which $(n, r; 1, G)$ - difference matrices D and which positive integers m is mD a maximal difference matrix. Necessary conditions are that D be a maximal difference matrix and n does not divide m. These conditions are not sufficient in general. Find sufficient conditions.

Problem 59. Find new parameter pairs (n, r) for which there exists a maximal r-clique of a group of order n. Such a parameter pair (n, r) implies the existence of a maximal set of $r + 1$ mutually orthogonal Latin squares of order n.

Problem 60. In Theorem 5.16 we established the maximality of certain cliques of $\mathfrak{C}_2(p)$ in Orth(Z_p), p an odd prime. For which odd prime powers does this result hold?

We have already shown that the result of Theorem 5.16 does not apply if $q = 9$. Using a computer, Pott (Preprint) has shown it to hold for $q = 25, 27$, and 49.

Problem 61. Find maximal cliques of $\mathfrak{C}_e(q)$, $e \neq 2$, that are maximal in Orth($GF(q)^+$). Some computer generated results for q prime and $e \neq 2$ are given in Pott (Preprint).

Problem 62. Is there only one $(p - 2)$-clique in Orth(Z_p), p a prime?

Problems 63 and 64 are the smallest interesting cases of problem 62.

Problem 63. Prove theoretically that there is only one 9-clique in Orth(Z_{11}).

Problem 64. Prove theoretically that there is only one 11-clique in Orth(Z_{13}).

Problem 65. Is there only one 15-clique in Orth(Z_{17})?

Problem 66. In Corollary 5.21 it is proved that the only $(p - 2)$ - clique of $\mathfrak{C}_2(p)$ is $\mathfrak{C}_1(p)$ if p is a prime. Is $\mathfrak{C}_1(p)$ the only $(p - 2)$ - clique of $\mathfrak{C}_e(p)$, e a proper divisor of $p - 1$?

Problem 67. In Corollary 5.22 the $(q - 2)$ - cliques of $\mathfrak{C}_2(q)$ are characterized for q an odd prime power. Characterize the $(q - 2)$ - cliques of $\mathfrak{C}_e(q)$, e a proper divisor of $q - 1$.

Problem 68. Theorem 5.21 establishes the non-existence of many $GH(n, Z_p)$ for $p = 3, 5$, or 7. Try to improve these results. We might for instance try to extend these results to the case $p = 11$.

Problem 69. In Theorem 5.22 G is assumed to be abelian. Does Theorem 5.22 still hold if this assumption is dropped?

Problem 70. If the Sylow 3-subgroup of G is non-trivial and cyclic must $\mathcal{S}(G) = \emptyset$? This is true if $|G|$ is odd as there must then exist a homomorphism from G onto its Sylow 3-subgroup. If $|G|$ is even then any Sylow 2-subgroup of G would have to be non-cyclic.

Problem 71. If G is abelian and its Sylow 2 and Sylow 3-subgroups are each either trivial or non-cyclic does G admit strong complete mappings? To prove or disprove this it is sufficient to restrict our attention to groups of the form $Z_3 \times Z_q$, $q = 3^n$.

Problem 72. Suppose that we are given a property P of (left) neofields. P might be the commutative property, right inverse property, left inverse property, inverse property, or exchange inverse property. Which groups can be the multiplicative group of a commutative (left) neofield, with property P, in which $1 + 1 = 0$? Which groups can be the multiplicative group of a commutative (left) neofield, with property P, in which $1 + 1 \neq 0$?

Problem 73. Prove the Hall-Paige conjecture for new classes of groups, in particular for new classes of simple groups, quasisimple groups, and almost simple groups.

Problem 74. Prove the HP-system conjecture for more classes of almost simple groups.

References

Afsarinejad, K. (1987). On Mutually Orthogonal Knut Vik Designs. Statistics and Probability Letters **5**, 323-324. [MR 88i:62138]

Anderson, B. A. (1984). Sequencings and houses. Congr. Numer. **43**, 23-43. [MR 86d: 05020]

Bailey, R. A. and Jungnickel, D. (1990). Translation nets and fixed-point-free group automorphisms. J. Combin. Theory Ser. A 55, no. 1, 1-13.

Bateman, P.T. (1950). A remark on infinite groups. Amer. Math. Monthly **57**, 623-624. [MR 12(1951),p.670]

Baumert, L. and Hall, M.Jr. (1973). Nonexistence of certain planes of order 10 and 12. J. Combinatorial Theory (A) **14**, 273-280. [MR 47(1974)#3206]
Reprinted in Evans (To appear - a)

Beth, T., Jungnickel, D., and Lenz, H. (1985). Design theory. Wissenschaftsverlag, Mannheim. [MR 86j:05026]

Bose, R.C., Chakravarti, I.M., and Knuth, D.E. (1960). On methods of constructing sets of mutually orthogonal latin squares using a computer I. Technometrics **2**, 507-516. [MR 23(1962)#A3099]
Reprinted in Evans (To appear - a)

Brock, B. W. (1988). Hermitian congruence and the existence and completion of generalized Hadamard matrices. J. Comb. Theory Ser. A **49**, 233 - 261.

Bruck, R.H. (1951). Finite nets.I. numerical invariants. Canadian J. Math. **3**, 94-107. [MR 12(1951), p.580]

_____ (1963). Finite nets.II, uniqueness and imbedding. Pacific J. Math. **13**, 421 - 457. [MR 27(1964)#4768]

Carlitz, L. (1953). A note on abelian groups. Proc. Amer. Math. Soc. **4**, 937-938. [MR 15(1954), p. 503]

_____ (1960). A theorem on permutations in a finite field. Proc. Amer. Math. Soc. **11**, 456 - 459.

Cates, M.L. and Killgrove, R.B. (1981). One-directional translation planes of order 13. Congr.Numer.**32**, 173-180. [MR 84h:51009]

Chang, L.Q., Hsiang, K., and Tai, S. (1964). Congruent mappings and congruence classes of orthomorphisms of groups. Acta Math. Sinica **14**, 747-756. (chinese).
Translated as: Chinese Math. Acta **6** (1965), 141-152. [MR 31(1966)#1220]

Chang, L.Q. and Tai, S.S. (1964). On the orthogonal relations among orthomorphisms of non-commutative groups of small orders. Acta Math. Sinica **14**, 471-480. (chinese).
Translated as: Chinese Math. Acta **5** (1964), 506-515. [MR 30(1965)#4690]

De Launey, W. (1984a). On the nonexistence of generalised Hadamard matrices. J. Statist. Plann. Inference **10**, no. 3, 385-396. [MR 85k:05027]
Reprinted in Evans (To appear - a)

_____ (1984b). On the non-existence of certain generalised weighing matrices. Ars Combinatoria **17A**, 117 - 132.

_____ (1986). A Survey of Generalised Hadamard Matrices and Difference Matrices $D(k, \lambda; G)$ with Large k. Utilitas Mathematica **30**, 5 - 29.

_____(1987). On difference matrices, transversal designs, resolvable transversal designs and large sets of mutually orthogonal F-squares. Journal of Statist. Plann. Inference **16**, no. 1, 107 - 125.

Dembowski, P. (1968). Finite geometries. Springer-Verlag, Berlin-Heidelberg-New York. [MR 38#1597]

Dénes, J. and Keedwell, A.D. (1974). Latin squares and their applications. Academic Press, New York-London. [MR 50(1975)#4338]

_____(1989). A new conjecture concerning admissibility of groups. European J. Combinatorics **10**, 171 - 174.

_____ (1991). Latin squares: New developments in the theory and applications. Annals of Discrete Mathematics **46**. North Holland, Amsterdam-New York-Oxford-Tokyo.

Dickson, L. E. (1897). The analytic representation of substitutions on a power of a prime number of letters with a discussion of the linear group. Ann. of Math. **18**, 65 - 120, 161 - 183.

Dinitz, J. and Stinson, D. (to appear). Room squares and related designs. In Contemporary design theory: A collection of surveys, edited by J. H. Dinitz and D. R. Stinson.

Di Vincenzo, O. M. (1989). On the existence of complete mappings of finite groups. Rend. Mat. Appl. (7) **9**, no. 2, 189 - 198.

Drake, D. A. (1977). Maximal sets of Latin squares and partial transversals. J. Stat. Planning Inf. **1**, no. 2, 143 - 149. [MR 58#5272]

_____. (1979). Partial λ-geometries and generalized Hadamard matrices over groups. Canad. J. Math. **31**(3), 617-627. [MR 81h#05031]
Reprinted in Evans (To appear - a)

Erdös, P. and Evans, A. B. (1989). Representations of graphs and orthogonal Latin square graphs. J. Graph Theory **13**, no. 5, 593-595.

Euler, L. (1779). Recherche sur une nouvelle espèce de quarrés magiques. [Memoir presented to the Academy of Sciences of St. Petersburg on 8th. march, 1779]. Leonardi Euleri Opera Omnia, Série 1, **7**(1923), 291-392.

Evans, A.B. (1987a). Orthomorphisms of Z_p. Discrete Math. **64**, 147-156.
[MR 88k: 05037]
Reprinted in Evans (To appear - a)

_____ (1987b). Generating orthomorphisms of $GF(q)^+$. Discrete Math. **63**, 21-26.
[MR 88f: 200501]
Reprinted in Evans (To appear - a)

_____ (1988). Orthomorphism graphs of Z_p. Ars Combinatoria **25B**, 141-152.
[MR 89e:05104]

_____ (1989a). On planes of prime order with translations and homologies. Journal of Geometry **34**, 36-41. [MR 90e:51015]

_____ (1989b). Orthomorphisms of groups. Annals of the New York Academy of Sciences **555**, 187-191.

_____ (1989c). Orthomorphisms of $GF(q)^+$. Ars Combinatoria **27**, 121-132.
[MR 90c:05039]

_____ (1989d). Orthomorphism graphs of groups. Journal of Geometry **35**, 66-74.

_____ (1990). On strong complete mappings, Congr. Numer. **70**, 241 - 248.

_____ (1991). Maximal sets of mutually orthogonal Latin squares I. Europ. J. Combinatorics **12**, 477 - 482.

_____ (Editor) (To appear - a). Advances in Discrete Mathematics and Computer Science. Volume VI. "Difference matrices, generalized Hadamard matrices and orthomorphism graphs of groups". Hadronic Press, Nonantum, Massachusetts.

_____ (To appear - b). Maximal sets of mutually orthogonal Latin squares II. European journal of Combinatorics.

_____ (To appear - c). Mutually orthogonal Latin squares based on linear groups. Proc. Marshall Hall Conf., Burlington, Vermont, September 1990.

Evans, A. B., Fricke, G. H., Maneri, C. C., McKee, T. A., and Perkel, M. (Preprint). Representations of graphs modulo n.

Evans, A.B. and Mcfarland, R.L. (1984). Planes of prime order with translations. Congr. Numer. **44**, 41-46. [MR 86d:51007] Reprinted in Hsu (1987), 291-296.

Fleisher, E. (1934). Eulerian squares. Ph.d. thesis.

Hachenberger, D. and Jungnickel, D. (1990). Bruck nets with a transitive direction. Geom. Dedicata **36**, no. 2 - 3, 287 - 313.

Hall, M. (1952). A combinatorial problem on abelian groups. Proc. American mathematical society **3**, 584-587. [MR 14(1953), p.350] Reprinted in Hsu (1987), 13-16.

Hall, M. and Paige, L.J. (1955). Complete mappings of finite groups. Pacific J. Math. **5**, 541-549. [MR 18(1957), p.109] Reprinted in Hsu (1987), 17-25.

Hedayat, A. (1977). A complete solution to the existence and non-existence of Knut Vic designs and orthogonal Knut Vic designs. J. Comb. Theory (A) **22**, no.3, 331-337. [MR 55#12548]

Hedayat, A. and Federer, W.T. (1969). An application of group theory to the existence and nonexistence of orthogonal Latin squares. Biometrika **56**, 547-551. [MR 41#6373] _____ (1975). On the non-existence of Knut Vic designs for all even orders. Ann. Statist. **3**, 445-447. [MR 51#4577]

Horton, J. D. (1990). Orthogonal starters in finite abelian groups. Discrete Math **79**, no. 3, 265 - 278.

Hsu, D.F. (1980). Cyclic neofields and combinatorial designs. Springer-Verlag, Lecture Notes in Mathematics, no. 824. [MR 84g#05001] _____ (Editor) (1985). Advances in discrete mathematics and computer science. Volume I. "Neofields and combinatorial designs". Hadronic press, Nonantum, Massachusetts. [MR 86f:05002] _____ (Editor) (1987). Advances in discrete mathematics and computer science. Volume II. "Generalized complete mappings". Hadronic press, Nonantum, Massachusetts. _____ (1991). Orthomorphisms and near orthomorphisms. In Graph theory, combinatorics, and applications, edited by Y. Alavi, G. Chartrand, O. R. Oellermann, and A. J. Schwenk, Wiley, 667 - 679.

Hsu, D. F. and Keedwell, A. D. (1984). Generalized complete mappings, neofields, sequenceable groups and block designs.I. Pacific J. Math. **111**(2), 317-332. [MR 85m#20031] Reprinted in Hsu (1985), 381-396.

_____ (1985). Generalized complete mappings, neofields, sequenceable groups and block designs.II. Pacific J. Math. **117**(2), 291 - 312. [MR 86k:05034]
Reprinted in Hsu (1987), 321-342.

Huppert, B. (1967). Endliche Gruppen I. Springer-Verlag. Berlin-Heidelberg-New York. [MR 37#302]

Hurwitz, A. (1882). _____. Nouv. Ann. Serie **31**, 389.

Johnson, C.P. (1981). Constructions of neofields and right neofields. Ph.d. thesis, Emory University.

Johnson, D.M., Dulmage, A.L. and Mendelsohn, N.S. (1961). Orthomorphisms of groups and orthogonal latin squares.I. Canad. J. Math. **13**, 356-372. [MR 23(1962)#A1544]
Reprinted in Hsu (1987), 29-45.

Jungnickel, D. (1978). On regular sets of latin squares. Problemes combinatoires et theorie des graphes (colloq. Internat. CNRS, Univ. Orsay, Orsay, 1976). Colloq. Internat. CNRS **260**, 255-256. [MR 80j#05024]

_____ (1979). On difference matrices, resolvable transversal designs and generalized Hadamard matrices. Math. Z. **167**, no.1, 49-60. [MR 81k:05032]
Reprinted in Evans (To appear - a)

_____ (1980). On difference matrices and regular latin squares. Abh. Math. Sem. Univ. Hamburg **50**, 219-231. [MR 81m:05034]
Reprinted in Hsu (1987), 155-167.

_____ (1981). Einige neue Differenzenmatrizen. Mitt. Math. Sem. Giessen, no.149, 47-57. [MR 82i#05016]
Reprinted in Evans (To appear - a)

_____ (1989a). Existence results for translation nets II. J. Algebra **122**, 288 - 298.

_____ (1989b). Partial spreads over Z_q. Linear Algebra Appl. **114**, 95 - 102.

_____ (1990). Latin squares, their geometries and their groups. A survey. In Coding Theory and design Theory, part II, IMA Vol. Math. Appl. **21**, Springer, New York-Berlin, 166 - 225.

Jungnickel, D. and Grams, G. (1986). Maximal difference matrices of order ≤ 10". Discrete Math. **58**, no. 2, 199-203. [MR 87e:05030]
Reprinted in Evans (To appear - a)

Keedwell, A.D. (1966). On orthogonal latin squares and a class of neofields. Rend. Mat. e Appl. (5)**25**, 519-561. [MR 36 (1968)#3664]
Reprinted in Hsu (1985), 96-138.
Erratum MR 37(1969), p.1469.

_____ (1983). The existence of pathological left neofields. Ars Combinatoria **16B**, 161 - 170. [MR 85k:20203]

Lidl, R. and Niederreiter, H. (1983). Finite fields. Encyclopedia of mathematics and its applications, vol.20. Addison-Wesley, Massachusetts. [MR 86c:11106]

Lindner, C. C., Mendelsohn, E., Mendelsohn, N. S., and Wolk, B. (1979). Orthogonal latin square graphs. J. Graph Theory **3**, 325 - 338. [MR 80k:05022]

Mann, H.B. (1942). The construction of orthogonal latin squares. Ann. Math. Statistics **13**, 418-423. [MR 4(1943), p.184] Reprinted in Hsu (1987), 1-6.

_____ (1944). On orthogonal latin squares. Bull. Amer. Math. Soc. **50**, 249-257. [MR 6(1945), p. 14]

Mendelsohn, N.S. and Wolk, B. (1985). A search for a nondesarguesian plane of prime order. Lecture notes in pure and applied mathematics, volume 103. Marcel Dekker, New York, 199-208. [MR 87g:51014]
Reprinted in Evans (To appear - a)

Mills, W.H. (1977). Some mutually orthogonal latin squares. Congr. Numer. **19**, 473-487. [MR 58#286]
Reprinted in Evans (To appear - a)

Niederreiter, H. and Robinson, K.H. (1982). Complete mappings of finite fields. J. Austral. Math. Soc. Ser. A **33**, 197-212. [MR 83j#12015]
Reprinted in Hsu (1987), 200-215.

Ostrom, T. G. (1966). Replaceable nets, net collineations, and net extensions. Canad. J. Math. **18**, 666 - 672. [MR 35(1968)#4809]

Paige, L.J. (1947a). Ph.d. dissertation, University of Wisconsin.

_____ (1947b). A note on finite abelian groups. Bull. Amer. Math. Soc. **53**, 590-593. [MR 9(1948), p.6]

_____ (1949). Neofields. Duke Math. J. **16**, 39-60. [MR 10(1949), p.430]

_____ (1951). Complete mappings of finite groups. Pacific J. Math. **1**, 111-116. [MR 13(1952), p.203]
Reprinted in Hsu (1987), 7-12.

Pott, A. (Preprint). Maximal difference matrices of order q.

Repphun, K. (1965). Geometrische Eigenschaften vollständiger Orthomorphismsysteme von Gruppen. Math. Z. **89**, 206-212. [MR 33#631]

Roberts Jr., C. E. (Preprint). Five mutually orthogonal Latin squares of order 24.

Roth, R. and Peters, M. (1987). Four pairwise orthogonal Latin squares of order 24. J. Comb. Th. Ser. A **44**, 152-155. [MR 88b: 05032]

Sade, A. (1963). Isotopies d'un groupoide avec son conjoint. Rend. Circ. Mat. Palermo (2) **12**, 357-381. [MR 29#4832]

Saeli, D. (1989). Complete mappings and difference ratio in double loops (Italian. English summary). Riv. Mat. Univ. Parma (4) **15**, 111 - 117.

Schellenberg, P.J., Van Rees, G.H.J. and Vanstone, S.A. (1978). Four pairwise orthogonal latin squares of order 15. Ars Combinatoria **6**, 141-150. [MR 80c:05038]
Reprinted in Evans (To appear - a)

Singer, J. (1960). A class of groups associated with Latin squares. Amer. Math. Monthly **67**, 235-240. [MR 23(1962)#A1542]

Sprague, A. P. (1982). Translation nets. Mitt. Math. Sem. Giessen **157**, 46 - 68.

Storer, T. (1967). Cyclotomy and Difference Sets. Markham, Chicago. [MR 36#128]

Studnicka, I. (1972). Non-existence of Cartesian groups of order $2p^m$. Comment. Math. Univ. Carolinae **13**, no. 4, 721-725. [MR 47#7589]

Vijayaraghavan, T. and Chowla, S. (1948). On complete residue sets. Quart. J. Math. (Oxford Ser.) **19**, 193-199. [MR 1949, p. 433]

Wan, D. (1986). On a problem of Niederreiter and Robinson. J. Austral. Math. Soc. Ser. A. **41**, 336-338. [MR 87k:11137]

Woodcock, C.F. (1986). On Orthogonal Latin Squares. J. Comb. Th. Ser. A **43**, 146-148. [MR 87j:05044]
Reprinted in Evans (To appear - a)

Index